Design for Innovative Technology

Design for Innovative Technology

From Disruption to Acceptance

Yves Mirande & Nicolas Henchoz

EPFL Press
A Swiss academic publisher distributed by Routledge

Contents

7 Foreword

Part One
A History, a Context, an Encounter

13 C’est Design !

25 *Sunny Memories*

39 Transdisciplinarity: Uniting Differences
44 Michel Maffesoli: Sensitivity and Reason

49 *Biomimesis*

57 The Birthplace of an Adventure
64 Pierre Keller: Beyond Function

69 *Give Me More*

83 Design and Technology: The Encounter
88 Patrick Aebischer: Opening up New Fields

93 *Hidden Carbon*

103 London, Paris, San Francisco

107 *The Colors of Dignity*

Part Two
Visions and Propositions

117 A New Digital Revolution
124 Andrea Branzi: A World Made of Exceptions

129 *Lazy Bytes*

141 Original and Super Normal
146 Jasper Morrison: Living with the Object

151 *Montreux Jazz Heritage Lab*

161 Towards the First World
168 Nadia Bruschweiler-Stern: Disintegrated Adults

173 *Under Pressure*

183 From the Idea to the Product
190 Yves Béhar: Reinventing the Design Business

195 *Seeds of Colors*

203 And Now…?

Appendix

209 Endnotes
211 Bibliography
213 Special Thanks
215 Authors

Foreword

The aim of this particular design book is to develop this discipline's role in the wider world of innovation, and the world of disruptive innovation in particular. *Disruptive* in this case means *a radical break*. In other words, innovation that is liable to foster radically different ideas and therefore breaks with conventional ideas. This book is based on the experiences of the École polytechnique fédérale de Lausanne's EPFL+ECAL Lab. Created in 2007 with the cooperation of ECAL (University of Art and Design, Lausanne), this laboratory aims to explore the potential of emerging technologies through design. It does so by proposing novel prospects for the use of these technologies, integrating them into a social and cultural context and formulating principles of expression, thereby turning them into unique experiences for users. It also addresses the challenges posed by technology, most notably the relationship between the physical and digital worlds. Our discussion is illustrated with projects from the Lab, and also looks back at its origins and sources of inspiration to help the reader grasp this merger between design and engineering. We have been fortunate to be able to draw on other initiatives that have preceded it – from the Ulm School, the MIT Media Lab to design thinking.

The book goes further still by attempting to put the EPFL+ECAL Lab's approach into a broader context. It cites major societal movements, major trends in design and the roots of our perception and, in this way, builds new bridges and provides insight to enable the fostering of new viewpoints, introduce new issues and encourage other disciplines (the hard sciences and the humanities) to continue the debate. But make no mistake about it – *this book is rooted in design*. It does not pretend to play a role in sociological debate, nor does it seek to act as an arbitrator between the tenets of postmodernity (like Michel Maffesoli, whose ideas greatly contributed to this book) and their critics, or even promoters of a new model (like Nicolas Bourriaud and his altermodernity). It does *make reference* to postmodernity, but only to better consider the major conceptual shifts that have served as indicators for the practice of design. Such shifts highlight issues related to innovation and invite us to take a step back with regard to our preconceptions of them.

It starts with a critical look at the perpetual refrain of a world in constant acceleration. Yes, information circulates at an unprecedented speed. Yes, globalization creates new pressure on productivity due to income disparity and social conditions. Yes, we have never produced and destroyed so much so quickly. But is the world really moving faster? Take the life of someone born in 1880 and who died in 1960 (80 years by analogy to life expectancy in the more affluent Western countries). What did this person experience in their lifetime? The development of collective transportation (train, subway and airplane), the birth of individual transportation (motorbikes and cars). The revolution in real time communication (telephone, radio and television). All of this, not to mention radical changes in our relationship to energy (refrigerators, freezers, central heating, air conditioning, etc.) and medicine (vaccines and antibiotics). Humankind had likewise started its conquest of the infinitely great (by leaving the Earth to travel into space) and the infinitely small (to understand the physical basis of matter).

Then, what has so fundamentally changed since the 1960s? There is no doubt that digital technology and its mobile variations have profoundly changed our daily lives and habits. But the web has been around for more than 20 years now (the age of the majority). And let's not forget, the Internet is based on network concepts that are more than 60 years old and has existed as a communication protocol since 1973. But, as the book suggests, digital technology nevertheless maintains a youthful relationship with society and its turbulent, contradictory, sometimes obscure character. Hence, it seems important to assert that, despite the excitement that agitates our lives, not everything is changing as fast as we think. We still need to spend time thinking about and renewing the relationship between technology, innovation and society – efforts can prove particularly useful for individuals and profitable for the business world. We hope that the proposals in this book raise new questions, open new perspectives and inspire other disciplines to take up the debate, uniting all the skills necessary for such a challenge.

Yves Mirande and Nicolas Henchoz, December 3, 2013

Part One
A History, a Context, an Encounter

This first part of this book sets the stage for adventure, diving gaily into the history of design and certain concepts borrowed from other disciplines like sociology. It also discusses the roots of the EPFL+ECAL Lab. Created in 2007, this lab probably would not have come into being without the unique developmental paths of the two institutions with which it is associated – the Ecole polytechnique fédérale de Lausanne (of which it is a part) and the University of Art and Design, Lausanne (its founding partner) – and the men and women who have made the adventure possible.

C’est Design!

Many gadgets are marketed as being "design," which is meant to convince the customer more than anything else. This does not make the job any easier for designers, who consider their contributions to be at the heart of the innovation process and social change. Without attempting to compete with leading design historians and thinkers, it nonetheless seems necessary to begin this book with a look at this discipline, its role and its context, before diving into the adventure of disruptive innovation. This form of innovation breaks with the existing, and is inextricably linked to the emergence of new technologies. The history of design is very much aligned with the major societal shifts of the last three centuries. More interestingly, design has played a very specific role in these shifts, as an expression of the underlying desires of a society. As with other areas of the art world, it acts as a sensor, a mirror, an actor of societal movements, albeit with one peculiarity: it can affect a very wide audience by its influence on mass production. This chapter is not an attempt to chronicle or catalogue iconic objects, but rather to contextualize their emergence, development, use and even application, through a few select movements from the 19th century to the present. It also aims to link these movements to the innovations of each era. This historical evocation focuses on the idea that this discipline is actually rooted in innovation – that it develops solutions to meet new demands, new market needs or the desires of a given society, but is also able to anticipate them. This legitimizes its role alongside technology laboratories. Design has been a catalyst for innovation since its emergence as a discipline in the mid-19th century. This role has constantly changed and adapted throughout its history. The following chapters will discuss its role in years to come.

Movements

In 1946, in London, the "Britain Can Make It" exhibition displayed a gutted bomber. Its "organs" were connected with wires to everyday objects, from a tractor to a pan. Here is the perfect metaphorical illustration of the role the war played in the birth of mass industrial design. The field experienced a boom just after World War II, and 1946 remains etched in the collective imagination as being a key year of development in the daily life of mass industrial design. This aftermath period indeed crystallizes reconstructive,

[1]

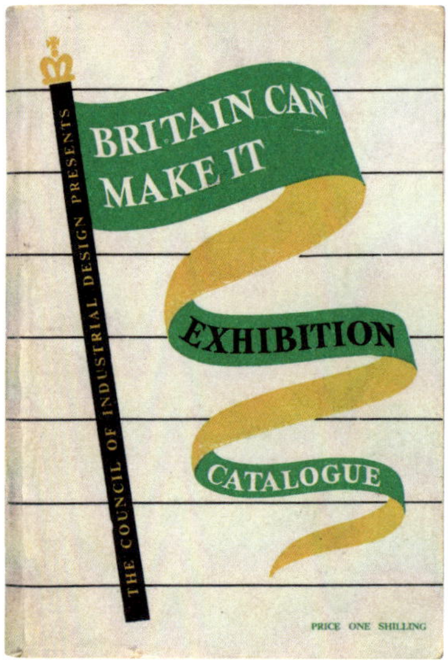

even curative energies, thanks in particular to the support the United States provided to Europe for its rebuilding in the form of the Marshall Plan (formally known as the European Recovery Program, ERP). Through a well-designed mechanism, this plan allocated funds to European states to allow them to make investments, among other things. Some companies thus were able to call upon artists, architects and designers to "*rendre belle la casserole*" – to quote sociologist Michel Maffesoli – in order to boost sales of their products. Design, however, was not simply born by spontaneous generation following the war; it is rooted in the history of our Western, even Eastern (and Japan in particular) societies, currents of thought and changes. Theoreticians agree that design was, in fact, born with the Industrial Revolution[1]. In her book *Design, Introduction à l'histoire d'une discipline*, Alexandra Midal writes, "It was in the organizational streamlining of the kitchen, conceived in the United States in 1841, that design has its origins." She goes on, but this time on the other side of the Atlantic: "But it was the Universal Exhibition in London in 1850 that launched design as a discipline in its own right. In the mid-19th century, the European Industrial Revolution pushed the British to consider the consequences of technical progress." In his book *Histoire du design de 1940 à nos jours*, design historian Raymond Guidot says, "The history of design is indissociable from that of the Industrial Revolution. With mechanized production began a new chapter of the human environment, written by industry and no longer just artisans and artists. From the infancy of the steam engine to the first skyscrapers, it develops by building on technological innovation." In *Design since 1945*, Peter Dorner also sheds light on this question: "Every technological revolution, every advance in the field of manufacturing seems to reserve a place for the designer." The Industrial Revolution[2] is therefore the veritable starting point of this discipline. This transformation would, for a long time, have a profound impact on agriculture, the economy, politics and the environment: in short, the whole of society. Hence, it is obvious that technical and technological progress and innovation are driving forces of design. Moreover, they were more or less welcomed by everyone, and had both critics and aficionados. These feelings were captured and reflected in the arts, especially by that which was called 'design' in France since the 1960s. Different movements thus have punctuated history with different currents, at times appropriating new technologies, at times criticizing them. But it is at the heart of this relationship and this tension that many innovations have emerged.

It all began with the nascent industrial world. All-powerful with its novelty, it did not burden itself with aesthetic creation. As Gilles Lipovetsky and Jean Serroy see it, "faced with the aesthetic

[2]

damage caused by the reign of modern machinery, two major currents of thought clashed." The Arts & Crafts movement[3], born in England in the 1860s, positioned itself against unbridled industrialization. John Ruskin, one of its pioneers, decried modern progress and its excessive mechanization (which would lead to the decline of society) in his book *The Stones of Venice* (1853). Ruskin also condemned the need for the individualization of production and expressed his concern about the rapid changes of societies under the impulse of the Industrial Revolution, which created a new social organization. His book had a tremendous impact, particularly through the idea of linking Art, Nature and... Man. By no means limited to England, this movement spread throughout the world, going by different names different countries ("Tiffany" in the United States, "Jugendstil" in Austria and Germany, "Nieuwe Kunst" in the Netherlands, "Modernismo" in Italy and even "Sapin Style" in Switzerland). In France, this movement came to be known as "Art Nouveau," and was characterized mainly by ornamentation inspired by trees, flowers, insects and animals, bringing sensitivity to everyday décor. It also rehabilitated the value of hand craftsmanship, creating a private realm considered favorable to personal development at the beginning of the 20th century. In France, its critics called it *style nouille* because of its characteristic arabesque forms[4]. This movement was clearly a manifesto against development, innovation and progress, emphasizing and promoting Nature as a bulwark against the encroaching Machine, using the style as a shield. The Art Deco movement was in much the same vein. Born in the 1910s, it reached its peak during the 20s before its decline in the 1930s. It too was inspired by nature, but in a much more stylized, geometrical way. Everything happened as though the rationalization of progress was just doing its job, intimating order in the figurative in order to rid itself of it. This movement was very much alive in architecture, in fashion and, of course, in "design."

In contrast to this anti-progress revolt, another major current of thought likewise has its roots in the 1850s, "inaugurated by Henry Cole, who gathered around him a group of reformist thinkers and artists." The novelty here, however, was not to reject progress and the machine, but rather to integrate them into a creative process "to invent a new language suited to the Industrial Revolution." In his book, Cole thus advocates for an "alliance between fine art and beauty and mechanized production." It was precisely this current that served as a base for the functionalist conceptions that endured up until the era of the Ulm School, and even later. "It already contains the seeds of what we will call industrial aesthetics or design to which the Bauhaus devoted itself in the 1920s." After training as a sculptor-engraver, Charles-Édouard Jeanneret (a.k.a Le Corbusier) traveled the world as a journeyman in search of a culture of architecture. In 1910, he served an apprenticeship in the studio of architect Peter Behrens in Berlin, where he met Walter Gropius and Mies Van der Rohe (also an apprentice). These encounters sealed intentions for the Modern Movement, which took shape during the 20s. "The first salon of the Union of Modern Artists in 1930, at the Marsan pavilion, responded to the presentation of an important cross-section of distinguished guests – the Deutscher Werkbund, the founders of the Bauhaus (Gropius, Breuer, Bayer, Moholy-Nagy) – and at the Grand Palais, in the artist-decorator salon that same year. These two events reflected the vitality of the Modern Movement, which presented itself as the only one capable of fulfilling the aspirations of industrial civilization." Among those involved in this movement – also called

Modernist – was Le Corbusier. The approach, stripped of all ornamentation and with clean, geometrical lines, emphasized rationality, thanks to new techniques and such materials as steel, glass and concrete.

After the proclamation of the Weimar Republic in November 1918, Gropius proposed uniting the school of decorative arts

[3]

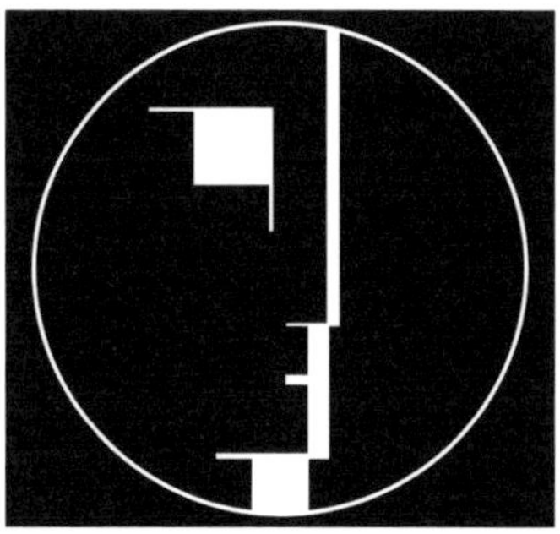

and the academy of fine arts of Weimar to the provisional government. On April 12, 1919, he was appointed director of the school, then called *Staatliches Bauhaus zu Weimar*. In the manifesto of the Bauhaus, Gropius describes the school's mission as follows: "The ultimate aim of all artistic activity is building!...Architects, sculptors, painters, we must all get back to craft, because there is no *professional art*!...There is no fundamental difference between the artist and craftsman...Let us together create the new building of the future which will be all in one: architecture and sculpture and painting." While Gropius originally wanted to introduce the idea of a return to craftsmanship, he did a 180° turn starting in 1922, foreshadowing the manner in which the arts would

[4]

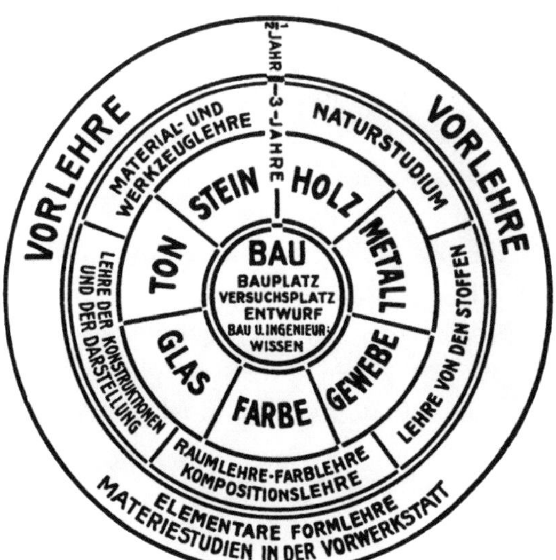

be taught and practiced throughout the 20th century. This shift is important because it marks the beginning of detachment from the idea of Nature in favor of the idea of Machine and Innovation. Ultimately, the Bauhaus' main idea was to bring art closer to the machine – which was forever more fascinating – following the precepts introduced in the 1890s by Louis H. Sullivan, who coined the famous phrase "Form follows function," and adopting the precepts of Adolf Loos, decreed in *Ornement et*

Crime.[5] Artist-designers like Marianne Brandt, Mies van der Rohe and Hin Bredendieck used materials from the industrial world, as did Marcel Breuer with his tubular steel furniture, the most well-known of which is the Wassily chair (or Model B3 chair). The Bauhaus was closed by the Nazis in 1933; its dissolution pronounced by Bauhaus leaders who found refuge in the United States. Ideas can be bullied and silenced but not die, and can resurface in a later era. Founded in 1955, the Hochschule für Gestaltung, or Ulm school, not only took up the aspirations of the teachings of the Bauhaus, but surpassed them in its ambition to go even further in rationalism, industrial production and the use of technological innovations. This ambition went hand in hand with the German post-war economic miracle. A good example of the school's application is without a doubt the collaboration of designers Hans Gugelot and Dieter Rams[6] with the firm Braun, which brought the company international recognition. Rams described his design approach as "Weniger, aber besser," which loosely translates as "Less, but better." Rams and his team created memorable products for Braun, including the famous SK-4 record player and the high-quality D-series record players (D45 and D46). However, the Ulm School's plan – as much political as functionalist– failed in some ways, as did the radical functionalism that it advocated, for the simple reason that it created objects that were too unrelated to life, resulting in a lack of appeal to the public. In short, these objects were too focused on function from which aesthetics had been entirely removed, and hence the emotional charge.

[5]

The Beginning of the Conjunction

We have just addressed – quite simply – the relationship between design and innovation, notably with the work of the firm Braun. Meanwhile in France, Roger Tallon, another great designer, was interested in the Ulm School. "Ulm is the school of thought where I discovered Gestalt Theory[7]...In the 50s, Ulm was the only place where anything was happening in terms of design, and I still think that what happened there overwhelmed everything else. But outrageous militancy hastened its decline...If Ulm had continued, it would probably have resulted in unequivocal formalization along the lines of the 'non-object,' with a complete absence of visibility," stated this designer who did so much for heavy industry (the TGV, the spiral staircase, Lip watches, and the Montmartre funicular in Paris) in the book that Chloé Braunstein-Kriegel and Gilles de Bure[8] devoted to his work. Roger Tallon[9] is a major figure in the relationship between design and innovation. After studying engineering (1944–1950), he joined Technès, a revolutionary technical and aesthetics consulting firm founded by Jacques Viénot, whose goal was to reinvent the relationship between engineering and design. Tallon, who was already teaching at the Ecole des Arts Appliqués in Paris, put Jacques Viénot's initiative in place: the first ever design course in France, under the name of Industrial Aesthetics Course. From this course was born the first generation of French industrial designers. The approach was decidedly in keeping with ambient modernism, but with a less dogmatic vision than the German approach. In 1963, he created the design department at the Ecole Nationale Superieure des Arts Décoratifs in Paris.

[6]

A New Rupture

In contrast to what could have been perceived as "functionalist dehumanization," the Italians meanwhile started their own little mass industrial design revolution. Italy was never completely industrialized, remaining largely agricultural. But starting in the 1950s, a shift took place – the "Italian miracle," to use Raymond Guidot's expression. Through a growing demand for amenities, coupled with cheap labor, Italian industry began to boom. It was at this point in time that partnerships between architect-designers and industrial designers were forged, resulting in iconic pieces like the Achille and Pier Giacomo Castiglioni 'Mezzadro' stool, for Nova Milanese in 1957. Its principle? A tractor seat (painted orange) screwed onto an L-shaped steel stem, stabilized by a piece of carved wood – a symbolic piece that combined strong and obvious cultural biases: the tractor seat (agriculture), the metal stem (industry) and carved wood (nature and craftwork) with a cheery, playful touch. Here we find the essence of Italian design heralding the beginnings of postmodernism. The advertising for the Lettera 22, a typewriter designed by Marcello Nizzoli for Olivetti in 1950, likewise exemplifies this new approach by drawing the device with a bouquet of flowers. Designer Ettore Sottsass created the Valentine typewriter for Olivetti in 1969, which also became an icon of design with its red cover. Italy also has a tradition of design thinkers – designer-architects who, despite working for famous firms, came together to distance themselves from their work and initiate strong manifestos, like Archizoom[10], which was created in 1964 by Andrea Branzi, Gilberto Corretti, Paulo Deganello and Massimo Morozzi. In 1967, Archizoom bisected a foam block to obtain pieces with random waves, thus creating the Superonda sofa. Randomness, gesture, bisecting symmetrical, parallel shapes are emblematic of the rational – a historic and symbolic attack on modernism. The wave design allows the user to utilize the object to sit in unconventional positions. Another example of this Italian revolution is Studio Alchimia, created in 1976 by Alessandro Guerrierro and Alessandro Mendini, who became its principal facilitator. For Mendini, inspiration came from shape. His research led him to undertake "redesign" operations (design procedures on existing, mundane objects such as the *Redesigned Thonet Chair,* 1978). In 1981, the Memphis group, launched by Ettore Sottsass, reacted to the post-Bauhaus designs of the 1970s, advocating for a sense of humor and derision, and investing objects with emotion. With the Italian approach,

[7]

[8]

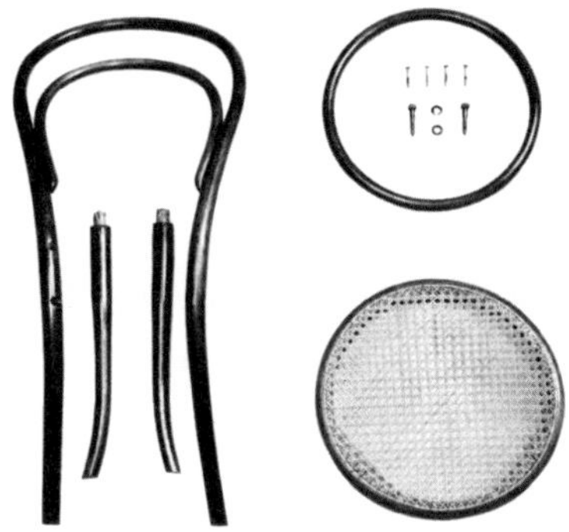

pop-culture references and dimensions of playfulness and empathy combined to renew the link with the user by making the latter its focus. This expression of postmodernism – a major current in design – corresponds to the emergence of postmodernity in society, which we will discuss in the next chapter.

Beyond Divisions

The rationalization of society by the Industrial Revolution thus brought about movements against this march toward progress, while leading others to surpass it through the expression of rationality itself, and ultimately turning it into dogma. Design therefore reveals both the hidden desires and anxieties of a society – consciously or unconsciously. It also maintains a strong link with the technological advances it then appropriates, lauds or refuses. Processed steel tubes allowed Marcel Breuer to make furniture. Later, plastic and even foam, led to the creation of such everyday items as ABS thermoforming displays (plastic) for Ettore Sottsass' Valentine (Olivetti 1969). The Industrial Revolution and technical innovation – these two dynamics are still at work today, and now include the digital world and its many revolutions, leading to new patterns of attraction and repulsion.

Given these amazing and inspiring movements that have made design what it is today, we would like, in this book, to go beyond the rational approach inherent to the emergence of technologies and to think outside of the box – no longer to choose sides but to couple, build bridges, and strive for transversality. Ettore Sottsass is a reference in this respect: a graduate of Politecnico di Torino (1939), he was nonetheless heavily influenced by his friend, painter Luigi Spazzapan, who offered him training different from that lavished upon him at school. Sottsass was able to reconnect with his initial desire to become a painter, and was particularly interested in color in his daily work. He said, "[for] me, design is a way to discuss life, society, politics, eroticism, food and even design itself. Ultimately, it is a way of building a potential figurative or metaphorical utopia of life. In any case, I don't think design should settle for giving form to more or less stupid products for an industry that is more or less sophisticated."

The EPFL+ECAL Lab ultimately has no other mission than to explore ways of discussing life and finding ways of integrating new technologies within it. For that is how invention becomes innovation. Other famous designers, like Philippe Starck and Yves Béhar, have tackled the subject. Several schools, like MIT's MediaLab, the Interaction Design Institute Ivrea in Italy and the d.school at Stanford, have launched initiatives. The EPFL+ECAL Lab is capitalizing on these past experiences, and the history of design itself, to propose a new model – rooted in innovation that appropriates technologies as they emerge from the laboratory, a model that calls for freedom of exploration and academic research, but with the aim of bringing it to fruition, and finally, a model that couples differences rather than reducing actors to the lowest common denominator. This is the thought revolution that will be discussed in this book.

[1] BRITAIN CAN MAKE IT
Exhibition Catalog, London, 1946.
© Design Council / University of Brighton Design Archive.

[2] STRAWBERRY THIEF
Drawing by William Morris, London, 1883.
© Victoria and Albert Museum, London.

[3] BAUHAUS LOGO
Created by Oskar Schlemmer, 1922.
© 2013 Estate Oskar Schlemmer, Munich.

[4] BAUHAUS EDUCATIVE SCHEME
Walter Gropius, 1922.
© Bauhaus-Archiv Berlin (BHA).

[5] ULM SCHOOL EXHIBITION IN THE SCHOOL'S CAFETERIA, 1958.
© Photo: Klaus Wille, HfG-Archiv Ulm.

[6] ROGER TALLON
Table, Module 400, 1965, cast aluminium and glass sheet, 185 × 105 × 72 cm.
Chairs, Module 400, 1965, polished aluminium and spazmolla, 64 × 40 × 40 cm.
© Photo : Marc Domage, courtesy of Jousse entreprise.

[7] VALENTINE
Olivetti typewriter by Ettore Sottsass, 1969.
© Associazione Archivio Storico Olivetti, Ivrea, Italy.

[8] N° 14 FROM THONET
Germany, 1859. First chair to be manufactured in series.
© Thonet GmbH, tous droits réservés.

Project

Sunny Memories

[1]

[2]
[3]

Berlin International Design Festival, 2011, the EPFL+ECAL Lab stand. Two design students create solar cells on a table before a stunned audience. The principal ingredients: two glass plates, a thin white paste and the juice of very ripe raspberries. Prepared right there on the spot in less than a minute, the cells are able to run a small electric motor. Further along, another table shows a more 'professional' version of these cells, which exhibit their colorful patterns, transparency and flexibility. On display are twenty or so objects that use these cells to revolutionize our relationship to solar energy. These cells now contribute to the visual expression of the objects, giving them unprecedented autonomy and allowing them to be incorporated into the most unlikely surfaces.

This revolution was born of 20 years of scientific research. In the early 90s, Professor Michael Graetzel and his team at the Ecole polytechnique fédérale de Lausanne designed a solar cell inspired by photosynthesis plant. At the heart of the process, a dye molecule turns light into electricity. The cell is therefore able to become transparent, colorful and flexible. Under the midday sun, it does not yet rival the performance of its more sophisticated traditional silicon competitors, with blue-violet metallic reflections. But when it comes to diffuse light (like the light in a house or on overcast day, for instance), the Graetzel cell provides amazing results. Most importantly, it aims at reducing the environmental impact and lowering production costs.

The first scientific display was wrought with a number of technical and scientific challenges, such as the long-term stability of the dye. While raspberry juice proved spectacular for the purposes of demonstration, it does not last long under the effect of the light. Creating strong, efficient cells, however, proved much more complex. But in a decade, research efforts have achieved certain milestones, and several industrial groups have acquired licenses to begin production. By the 2000s, Michael Graetzel had become one of the most quoted chemists of his generation, receiving some of the fields' highest distinctions (the Millenium European Innovation Prize, Prix Balzan and the Millennium Technology Prize). Yet, in terms of use, all was yet to be achieved. The earliest products were either inventions that were manufactured directly, such as translucent colored plates for architecture, or plates that were placed on the objects – an approach similar to that of traditional cells. It was simply a question of replacing the old technology [4] →

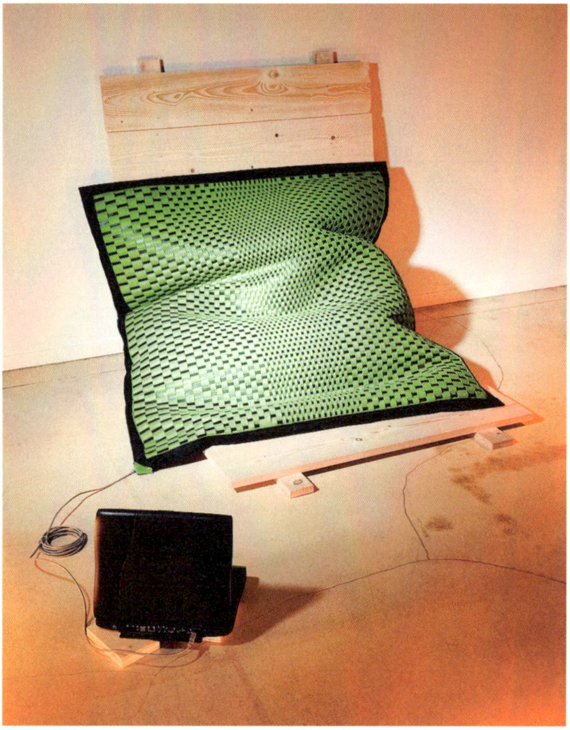

[5]
[6]

with the new in order to derive some cost- and performance-related benefits.

However, the introduction of color, flexibility and transparency can radically change our relationship to solar energy, given that the energy source was no longer limited to a plate that covers the product, but rather something that could become a fundamental part of the object itself. Hence, screen-printed patterns became a source of energy. This capacity for integration increased the types of surfaces for capturing light by reducing the importance of the performance criteria per square meter – especially given that a growing number of everyday objects have electronic components but do not require significant power.

For these objects, photovoltaics are not only an environmental advantage; they also provide autonomy, freeing them from dependency on batteries, chargers and cables. The project Sunny Memories was born of such thinking, proving that dye-sensitized solar cells not only replace silicon ones, but also opening up new prospects in terms of our relationship with objects that carry information and action, and consequently are often guardians of memory and heritage. By making them autonomous in terms of energy, we safeguard this memory. By a fortuitous set of circumstances, this theme converged with the vision of the private bank Lombard Odier & Cie in Geneva, which had just signed a partnership agreement with former Vice President Al Gore's company for responsible investment. Thierry Lombard, Managing Partner, wanted to embody this role of pioneer by actively participating in exploring the potential of these new solar cells.

The vision of solar energy proposed in this project was not exhaustive, but rather was a starting point for escaping classic representations of the use of traditional cells. Four design schools – the Royal College of Art in London, ENSCI-Les Ateliers in Paris, California College of the Arts in San Francisco and ECAL in Lausanne – agreed to participate in the project, inviting some of the biggest names in design as teachers (Yves Béhar, Durrell Bishop, Sam Hecht, Jörg Boner, and Jean-François Dingjian, to name only a few).

Michael Graetzel's lab opened its doors to students to aid in their understanding of how the cell was made, along with its challenges. Cell manufacturers like Solaronix offered their support to test the principles and make several prototypes. With its own

VOL +
PREV
NEXT
VOL –

team and partners, the EPFL+ECAL Lab was able to create a unique environment for the young designers, from the research laboratory to the end product. All in all, 88 students participated in the project, which lasted an entire semester – an unprecedented creative mass with very different approaches. At the Royal College of Art, the professors bought a few flimsy solar products in a large retail store, gave them to the students and said, "You've got a better solar cell, so do better!" After many presentations, meetings, discussions and trials, 28 projects were chosen to be included in the final exhibit; six were used as prototypes for use with real cells. Originally scheduled for four stops – Paris, London, Lausanne and San Francisco – the exhibit made ten stops, including the inauguration of The Laboratory at Harvard exhibition space, the American Institute for Architecture in NYC and the World Engineering Convention. Despite its long journey and the tired state of the objects, Sunny Memories was named among the ten best projects at the Berlin Design Festival in June 2011 – perhaps because the festival took place at exactly the same time as the raspberry harvest!

Beyond cultural and media success, what prospects does Sunny Memories have to offer? The lessons are many. To illustrate, we could cite just a few of the projects from among the 28 that were selected. For instance, the hat and scarf from the Dress Smart, Live Cool project. In addition to their playfulness, they reflect the complementarity of the design and engineering approaches. Michael Graetzel's lab spent years trying to improve the cell's performance. The very idea of covering it with cloth – thus limiting the sun's rays – is obviously counter-intuitive. More concerned with the user's experience than the technical performance, and above all completely unaware of the years of hard work to achieve its current level of performance, ECAL's Ismaël Studer and Virgile Thévoz made the leap, after discovering that a thin white linen cloth seemed to let through a lot of light. In the end, the cell lost only 15–20% of its efficiency. While this loss is not negligible, the cell remains functional and offers novel prospects in terms of appreciation, with its real cloth texture surface.

[7] Another example: Georges Moanack of the Royal College of Art offered a unique take on the flashlight by making it less fragile, heavy, and rigid – yet remaining autonomous – thus changing the object's identity and our relationship to it. Several other projects have also shed new light on the lamp concept, playing on the

[8]
[9]

[10]

idea of storing energy by day and emitting it by night. Incidentally, Solaronix developed real solar benches two years later, some of which were installed on the EPFL campus. The principle: these benches store up the sun's energy during the day and offer a halo of light that radically changes our perception of urban furniture or a park at night. Emile Rosen's Heliotone radio demonstrates how different layers of features can be combined: a thin transparent layer makes the surface of the solar cell tactile, turning it into an interface for adjusting the volume and channel. Another device makes this same cell vibrate, turning it into a speaker as well: thus the cell becomes a radio. Features, uses, visual languages, and materials – this project shows the extent to which design can unlock the potential of emerging technologies. Subsequently, Sunny Memories has given rise to several new projects with concrete industrial prospects including solar furniture.

CURATOR
Nicolas Henchoz

SCHOOLS AND INSTRUCTORS
California College of the Arts (CCA), San Francisco:
Yves Béhar, Rick Lewis
Royal College of Art (RCA), London:
Christophe Behling, André Klauser, Sam Hecht, Durrell Bishop
ENSCI-Les Ateliers, Paris:
Jean-François Dingjia, Eloi Chafaï
University of Art and Design, Lausanne (ECAL):
Jörg Boner

PROJECT ENGINEER
Fanny Riedo

SCIENTIFIC PARTNER
Laboratoire de photonique et interface EPFL,
Pr Michael Graetzel, Dr Paul Liska

SUPPORT
Lombard Odier Darier Hentsch

TECHNOLOGICAL PARTNER
Solaronix, G24i, Dyesol

SCENOGRAPHY
Big Game

PHOTOGRAPHY
Tonatiuh Ambrosetti, Daniela Droz

[1] HORN
Guillian Graves
ENSCI
Autonomous monophonic speaker plays digital music sources. The shape is reminiscent of a horn of a gramophone. It also optimizes the sound's amplification and the cells' efficiency in lower light environments.

[2] HELIO
Léa Longis
ENSCI
In an age where smaller is better and electronic devices are becoming lost in miniaturization, Helio makes its presence felt by making the radio the prominent object it was in the past. Reappropriating the look of tube amplifiers, the device is graced with a set of cells whose color and posiion complement one another to capture virtually the entire light spectrum.

[3] N47
Christophe Machet
Enrique Illanez
ECAL
A regenerating bowl for portable devices that stores all of our personal data – in other words, the fruit of our knowledge. Note the perpendicular solar cells on the surface, which play on its thickness. A novel and radical approach that offers great potential.

[4] FLOATING CHAIR
Diana Chang
CCA
This urban furniture for a public park emits a luminous halo at night, inviting passersby to have a seat, with a pleasant feeling of security. The lighting (which is located under the seat) makes it look as though the bench is floating. Its performance is effective thanks to its base, which is comprised of a continuous set of dyed solar cells.

[5] HANG
Georges Moanack
RCA
The same principle as a "Do Not Disturb" hotel sign, it becomes a flashlight. Its phosphorescent "night" side allows the user to locate the lamp and to personalize it for use in a child's room. The "day" side, made of flexible solar cells, serves to charge the lamp's batteries. A diode is located on one end of the object.

[6] ACEDIA
Geoffroy Buttey
Vanessa Hordies
ECAL
Because of its great flexibility and graphic potential, solar energy can be incorporated into a rather surprising range of objects. This pillow, braided with solar strips and felt, is the result of both visual and tactile exploration.

[7] HELIOTONE
Emile Rosen
RCA
In fact, this solar cell is an entire radio. An example of integration, a tactile layer allows it to act as an interface. The NXT SurfaceSound® actuator also turns it into a speaker.

[8] DRESS SMART, LIVE COOL
Ismaël Studer
Virgile Thévoz
ECAL
Fresh, light and cool – all adjectives that seemingly contradict the idea of photovoltaics. The designers challenged this paradox after seeking, at length, a cloth that could change the appearance of the cells without considerably reducing their effectiveness. The accessories then use the energy stored during the day to emit light in the dark.

[9] YOU'VE GOT MAIL
Clare Hsu
CCA
The ultimate link between the virtual world of email and the physical world of snail mail. The mailbox sensor alerts the owner by an SMS text or e-mail when physical letters, packages are deposited inside the mailbox. The touch privacy sensor and the communication system's electric supply come from the red solar cell integrated in the mailbox's lid.

[10] SUMMER NIGHT'S DREAM
Julia Roth
Véronique Baer
ECAL
A parasol by day, mood lighting at night, this project plays on the semi-transparency of solar cells. The principal of folding creates a morphological change between day and night.

Transdisciplinarity
Uniting Differences

Do not flee upon reading the title of this chapter. Behind these words lies a very simple idea as old as humanity itself: that of *going towards* – towards the unknown, towards something or someone we don't know. It is also the idea of gathering – gathering people, ideas, talent, convictions, and disciplines – and ensuring that these ideas, people, talents, and disciplines can be made to *agree*, in order to create something that works. Not a standardized system, but *something* – magma, a matrix, what have you. Transdisciplinarity is a scientific and intellectual position that aims to understand the complexity of the modern world and of our times, and lies simultaneously *between*, *through* and *beyond* all disciplines: the process of integrating and surpassing individual disciplines. The Charter of Transdisciplinarity says that "Transdisciplinarity complements disciplinary approaches. It occasions the emergence of new data and new interactions from out of the encounter between disciplines. It offers us a new vision of nature and reality. Transdisciplinarity does not strive for mastery of several disciplines but aims to open all disciplines to that which they share and to that which lies beyond them."[11] A complicated way to describe the revolution that is currently underway.

From Vertical to Horizontal

The need to organize "knowledge" (meaning a body of known facts or reproducible skills acquired through study or experience) has existed since the birth of knowledge itself. Not an easy task! And yet this organization began many millennia ago, as evidenced by the appearance of writing in Mesopotamia and Ancient Egypt. During the Middle Ages, apprentices gave knowledge a mythical, almost spiritual dimension. Starting in the 18th century, medicine started to organize knowledge based on a system of peer verification. This approach gradually spread to other disciplines – which were becoming increasingly numerous and specialized – until finally becoming an indispensable practice in the scientific world in the 20th century. These approaches reflect the "verticalization" of knowledge and its transmission in ever more specialized niches. The world of research – the cradle of new knowledge – is entirely structured on this verticality: funding, the validation of findings, value enhancement and recognition are done by groups linked to very specific disciplines.

Verticalization also affects the business world, which is organized into departments, as Henry Mintzberg demonstrates in his book *Mintzberg on Management: Inside Our Strange World of Organizations*. The pattern of thinking is thus established: we think vertically. In fact, we have done so ever since the "establishment" of the modern world. Not the one beginning in 1492 – the date historians claim as the beginning of Modern Times, but rather since rationality itself came into our thinking. In other words, since the 17th century, particularly with René Descartes' *Discourse on Method*. "To guide reason and seek the truth in the sciences" reads the subheading to the 1637 edition. With ideas come famous quotes. For instance, that men must "become masters and owners of nature" through advances in technology. In other words, dominate nature at all costs. Or Jean-Jacques Rousseau's *Social Contract*, published in 1762. This work was a turning point for modernity, emerging as one of the major texts of political philosophy by asserting the principle of the popular sovereignty. Make everything controllable, manage everything, compartmentalize it, put it into boxes and rationalize it. Such has been the abrasive work of modernity. This way of thinking, seeing and learning has its good points. Make everything straight and vertical so as not to scatter it. Dissect and isolate in order to better study. Yet starting in the 1950s, a new phenomenon no longer advocated those vertical systems to which we had grown so accustomed, but rather a way of seeing, arranging and thinking that is horizontal.

The Aftermodernity

In his work *La formation de l'esprit scientifique,* Gaston Bachelard urges us to "put scientific culture in a state of permanent mobilization, replace closed, static knowledge with knowledge that is open and dynamic...and to finally give reason a reason to evolve." But before trying to think like Gaston Bachelard, modernity and rationalism have done their work. In simplified terms, rationalism is a doctrine which claims that real knowledge results only from logical discourse and reasoning. Modernity draws inspiration from such thinking but is fueled by bigger ambitions: it has become a way of functioning for a society that is striving to free itself from the arbitrary, in order to become a well-organized, modernist society that seems to be the master of its own destiny.

Today, rationalism still remains at the heart of the production of scientific knowledge, but the modernity that brought it about has reached its saturation point. Science has not abandoned this rationalism that gives it its strength, but rather continues to break free of the modernist shackles that have partitioned it: nowadays, disciplines are organized more freely, forge links and build bridges. The movement for free access to scientific knowledge and cultural heritage works is supported by more than 400 universities around the world, which have signed the 2003 Berlin Declaration. Major academic institutions are making their teachings available and accessible to people around the globe by putting them online – the famous Massive Online Open Courses (MOOCs).

To describe the end of modernity, certain sociologists spoke of *post*modernity, followed by other concepts such as *hypermodern times* or *altermodernity*. The term postmodernity appeared in 1979 in Jean-François Lyautard's book *La condition postmoderne*. Here he describes the end of major thought systems, or the major explanatory systems like Marxism, culturalism, functionalism, etc. inherited from the 19th century.

In the second half of the 1960s, the revolution against modernity became more spectacular and protest-oriented, with "artistic" movements like Archizoom (1966–1974). Their authors did not hesitate to use icons of pop culture to violently lash out at the tenets of modernism, as reflected in such ideas as "superarchitecture" and "afro-tyrolien kitsch." The mosaic – the resulting visual patchwork – likewise attests to this idea of Trans-Something. This powerful notion of combining differences is the foundation of postmodernity.

Decidedly Innovation-oriented

In the spotlight since the 1990s, transdisciplinarity has brought together numerous disciplines. In Lausanne, the idea of a new laboratory was built around that of meeting: the idea of combining design and engineering approaches, which today demand radically different cultures and approaches, especially given design liberation from modernism over the past decades. Gone is the convergence that prevailed in Roger Tallon's day. Engineers focus on creating new knowledge and developing new scientific achievements. The building of scientific knowledge – with its factual rigor and peer verification processes – requires time and resources. When a major discovery is ultimately made, the researcher wants to develop it as quickly as possible. Competition does not wait! Performance may be enough when it comes to a component of a larger system, but for popular application, it's a different story.

The general public in the West has long distinguished between technology and progress. The EPFL+ECAL Lab thus calls upon design to explore the potential of emerging technologies, to give them purpose, meaning and makes them resonate with the social and cultural contexts of daily life. Hence, design is not a simple question of style, aesthetics or ergonomics. In fact, it is no longer a cosmetic issue but rather a key aspect of thinking about design. It reflects a form a pop culture because it is expressed here in its contemporary form and demands the strength of an "inspirational," sensitive and contextual approach. It takes advantage of the freedom inherent in the academic setting to explore without limits the opportunities and the technological results provided by rational construction. It is no longer a point of contention but rather is a productive richness at the heart of the innovation process – an essential, groundbreaking shift in ways of thinking. And it is indeed because these notions – horizontality, transversality, meeting and uniting differences – have existed for some time now that it was possible to create this laboratory.

Interview with Michel Maffesoli
Sensitivity and Reason

Michel Maffesoli loves to receive guests in his spectacular library, which is stacked to the ceiling with the major references of the human and social sciences, but also more unusual things, suggesting eroticism. With sparkling eyes, dapper bowtie and red shoelaces joyfully teasing the austerity of academic knowledge, Maffesoli is not only the sociologist of postmodernity, he is an incarnation of it. With mischief, humor and pleasure (and often roaring with laughter), he is constantly creating links between his field's development and popular issues. A former student of Gilbert Durand and Julien Freund, Maffesoli is a professor at the University of Paris Descartes. Holder of the Durkheim chair of sociology, an Officer of Arts and Letters and a Knight of the *Legion d'Honneur*, he has developed work around questions of community social ties, the prevalence of the imaginary and everyday life in contemporary societies. His research and thinking have greatly contributed to the spread of the Post-Modern model, thus encouraging the development of a comprehensive sociology. For him, the little things of everyday life, like discussions on street corners, are material for thinking about the social. For more than 40 years he has been distilling his enlightened (and enlightening) vision of societal life and carefully sharing this thinking through notable appearances in the media and, most notably, his many books which have become major references (*Le temps des tribus*, *L'ombre de Dionysos*, *Le réenchantement du monde* and most recently *Homo Eroticus*).

When exactly do you place this shift from modernity to postmodernity?

For me, on a visual level, the shift had already taken place in 1950. It was at this precise moment that we saw a real visual caesura. We can see it clearly, especially with Italian-American architect Robert Venturi, who fought against right-angled modern architecture, the Bauhaus, Le Corbusier, etc. Venturi distanced himself from this rational, minimalist expression and the *machine à habiter*. He wasn't afraid to use and mix a variety of styles – a little Baroque, a little Roman, a little Rococo – to create a mosaic of different styles. Venturi advocated the use of patchwork and mosaic. This was indeed the architectural birth certificate of postmodernity – not something that could be reduced to unity but rather diversity and roots.

The end of modernity doesn't seem obvious to everyone. How do you explain that?

There is effectively a delay on the part of the intelligentsia, meaning those who have the power to say and do (journalists, academics and politicians). This intelligentsia remains set on modern values, like the social contract. They don't realize that there is another way of relating to the earth, to the world and to others that is in gestation. One must understand that the adventure of the mind is elsewhere...

Postmodernity starting in the 1950s, postmodernism starting in the early 1970s (and not without some violence and radicalism)...in your opinion, how did this evolution segue?

You know, I'm no design expert, but I can tell you about the logic behind this journey. And that logic is when you have a little boost right at the beginning, in other words when you go and aestheticize functionality. But as you say (and you're right) this evolution is still very modern. But a rift appeared in functionality, and once there's a rift, invasion becomes possible. The fleet arrives, carrying postmodernism with it.

Technology, with all its rationality, is nonetheless at the heart of innovation. How can we bring it face to face with other forms of creativity, such as design?

In my book *Eloge de la raison sensible* (where I raise this question), I do not disassociate reason and sensitivity. Reason is the logic of technology. In *The Structure of Scientific Revolutions,* Thomas Kuhn showed that, starting in the 18th century, rationalism created scientific and technological development. It was the high performance of the Western European model. In design, we do not negate reason; we simply couple it with the sensitive.

In your opinion, how do we unite these different disciplines productively?

We have to let them be. We're used to always looking for solutions. Our default mode is always "find a solution." But the peculiarity of postmodernity, again, is "the mosaic," in which each element preserves its original configuration, color, form, etc. And yet, coherence emerges within it. It holds together. I often talk with business owners faced with the challenge of linking different disciplines in their projects. They see the limits of typical Taylorism, feel stuck in this type of rationalism, and yet must deal with a new generation that comes with its own irrational, creative dimension. Essentially, we could say that there is conflictual harmony. That would be my definition.

How to think about our relationship to the object today?

While the object was functionalized during the first half of the 20th century, postmodernity allowed for a revival of the magical relationship to the object that existed before the Modern period, which had been lost. For me, laptop computers are transitional objects. We have a magical relationship with them, just as primitive tribes had a magical relationship with their totems. I would call it magical and mystical participation: when my laptop isn't working, I insult it, I develop a conflictual or affective relationship with it. Sometimes I get furious with it, sometimes…it's not rational [laughs]. Everyone has a very symbiotic relationship with this type of object, built on the synergy of the archaic and technological development.

Remember "Inanimate objects, do you have a soul?" Well, I'd say yes, of course! Whereas before we had the objectified objects, today we have what I would call the idea of the journey.

We hear more and more about the notion of empathy, especially in design...

It's one of the concepts I've been interested in for a long time and which I have written about. Empathy is shared pathos. We participate: we realize that there is a back and forth, reversibility, feedback. This concept comes from Lucien Lévy-Bruhl, a 20th century-ethnologist who reflected and worked on the notion of primitive mentality. He was the one who used the expressions *magical participation* and *mystical participation* insofar as primitive tribes in general "participate" with their objects. These objects were a piece of the world, the cosmos, and so there was respect for them – not just the mere power of domination. They were earth, wood, stone, etc., and hence things from the great matrix of life. Modernity completely forgot this. With postmodernity, it once again seems important to say "I am part of [this matrix]." So we beautify, pay attention and protect, because the object – like me – once again is a bit of the world. And the result is another relationship to the world that is more participative. Basically, empathy is nothing more than that.

Project

Biomimesis

[1]

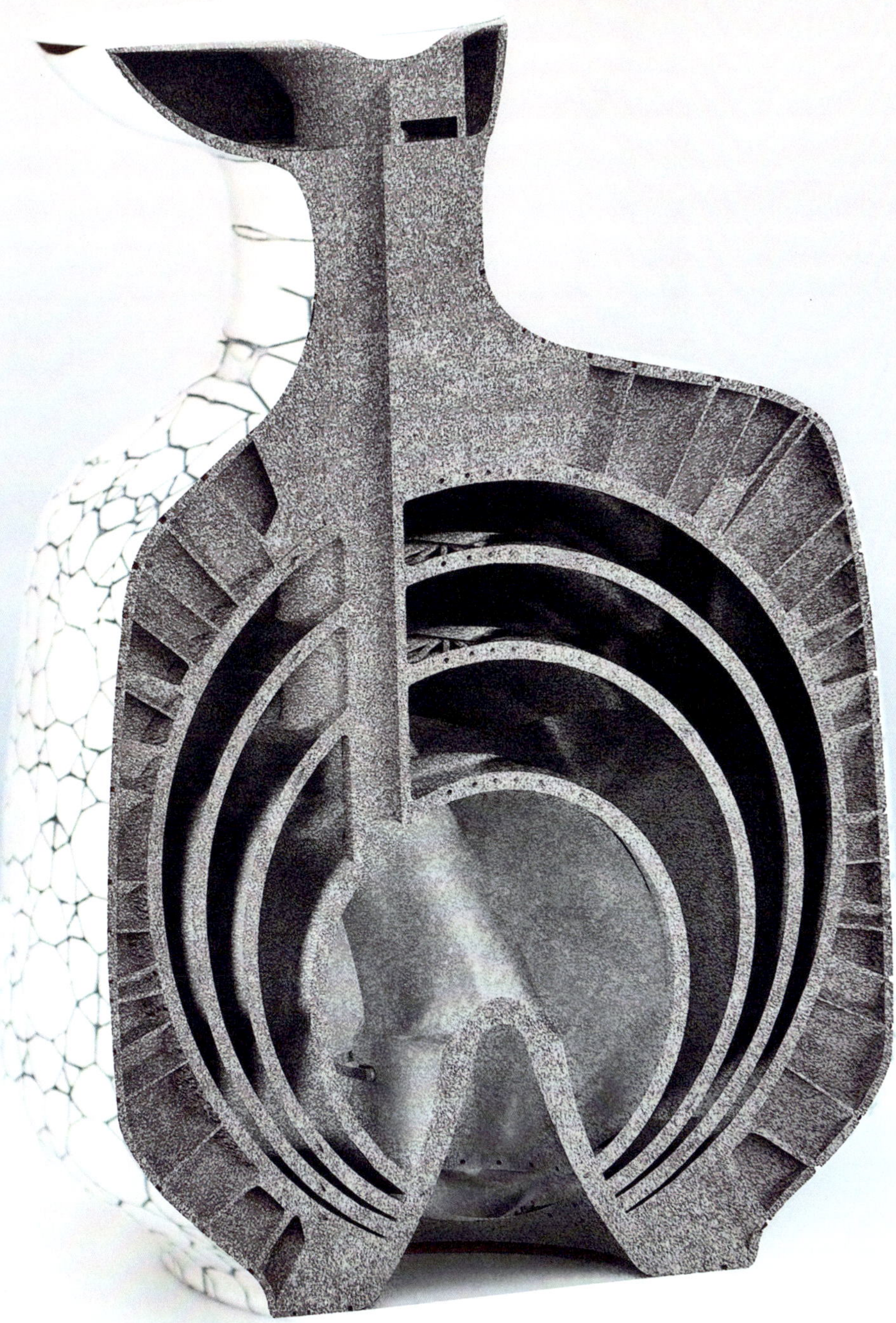

[2]
[3]

Like the atom in last century, bio-inspiration now appears to carry the certainty of a better world. At every level – from the vegetalized air purifier to the promise of transforming inert planets into inhabitable worlds – progress is definitely on the march! The EPFL+ECAL Lab is no exception when it comes to this topic. It brings together several institutions and has launched a dual masters program. One half of the pair of researchers is Michka Mélo, a science and technology student at the Ecole polytechnique fédérale de Lausanne, who wanted to devote his research to developing a methodology for biomimetic design. His aim was to re-examine the principles evoked by many authors – from Janine Benyus, founder of the discipline, to the latest studies from the Biomimicry Institute – with the rigor and logic of a polytechnique graduate. To do this, he drew upon the expertise of the University of Lausanne's Industrial Ecology Group. The other half of the pair is Guillian Graves, a student at the ENSCI-Les Ateliers in Paris, who wanted to address the same question but from an industrial designer's perspective. The duo's mission was to engage in an uncompromising relationship – not to create a chimera, but with the aim of both producing convincing work in their respective areas.

The first step was to establish a common core of knowledge and reference, to share a similar language and explain their respective approaches. They therefore had to go back to the basics: the definition of sustainable development from the 1987 Brundtland Report and the benchmark work 'Innovation Inspired by Nature' published by Janine Benyus in 1997, as well as a plethora of scientific articles and references in the field of design. The pair managed to set out a series of clear definitions, summarized here in these excerpts from Michka Mélo's report: "A bio-inspired system is any artificial system inspired by the living." Bionic thus corresponds to bio-inspired systems designed for the purposes of technological performance. Biomorphism means the imitation of living systems for aesthetic purposes, while "biomimicry refers to the conception of bio-inspired systems for sustainable development purposes."

The project gave rise to questions of principle: "Let's put ourselves in the shoes of an actor who wants to work for sustainable development, using the living as his or her inspiration. How do we find lasting solutions?" And even more fundamentally: "Is the living a model of sustainability?"

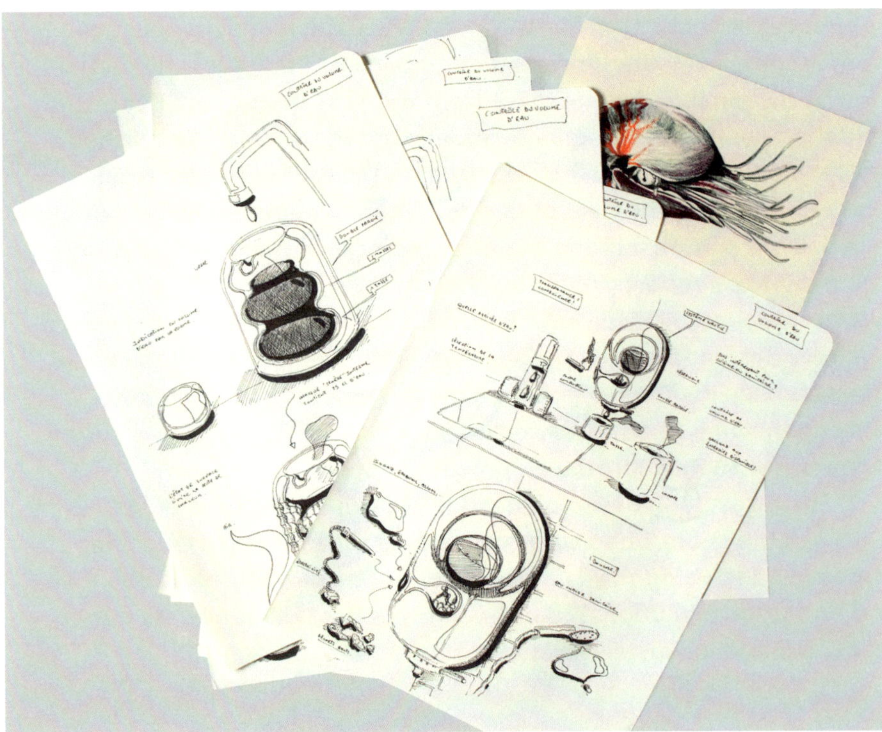

[4]
[5]

The partnership with Guillian Graves created a permanent dialogue between a theoretical approach and potential applications. The designer was responsible for developing tangible objects at the end of his study. "After our meeting we quickly decided to focus our research on the living environment. The intimate nature of this place also applies to our relationship with objects. We interact frequently – and even emotionally – with certain objects, and so we wanted to use this relationship to the object as a point of leverage to begin the transition towards a more sustainable lifestyle." Thus, very early on in the project, design played a role, as did engineering. The researchers chose an object to incarnate this approach: "The kettle, a very common object in our society, has a simple function: to produce hot water for food consumption, for example, for a hot drink or for cooking food. It is therefore at the crossroads of different flows of matter and energy in the kitchen, making it an excellent object to embody this approach." It also touches upon several issues, including energy consumption and recycling.

This concrete example served as an experimental field for the researchers to develop and test their overall design approach. The approach starts with a functional and environmental analysis of the chosen object, from which emerged specific design challenges. For the kettle, it is mainly the question of the volume of water to heat, the final temperature of the water, the insulation capacity and the source of energy. The second step is based on biological research. It aims to identify mechanisms in the living world that are likely to provide useful guidelines for addressing these challenges. For example, the chambered nautilus shell led to the idea of dividing the interior of the kettle into several compartments adapted for different uses. The toucan's bill led to the idea of creating an extremely insulating wall that is highly economical with regard to material. Finally, the temperature regulation in termite mounds (based on air flow) led to the optimization of the heating system. These principles then supply the next phase, known as concept generation, in which the concepts are confronted with the object's many limitations, most notably as relates to the context of use, technical feasibility, utilization, and their mutual compatibility. This eventually results in the final design and its evaluation. The final step consists in comparing the new product with existing products to measure the benefits obtained.

Michka Mélo – the engineer – formalized, structured, and organized the approach. He also conducted scientific and technological assessments (in-depth bibliographical research, calculations and laboratory experiments). Guillian Graves – the designer – sought solutions and incorporated them into a final concept. Two concrete projects emerged from this work: the "Nautilus" kettle and a full-service heating system with accessories.

Throughout their work together, the two researchers kept a laboratory journal in which they took turns describing the project's progress as well as any confusion and doubts they had. In a paragraph entitled "Feedback," Michka Mélo declared "It is [...] not easy to combine the methods of design and engineering, specifically because it is not easy to achieve the harmonious cohabitation of scientific reasoning's rigor and the freedom of qualitative elements. Nevertheless, we are both convinced that it is by doing more of this kind of experiment that we can meet challenges in where the quantitative and qualitative are interwoven, as is the case for sustainable development."

AUTHORS
Designer: Guillian Graves, ENSCI-Les Ateliers
Engineer: Michka Mélo, EPFL

SUPERVISION
Pr Alexandre Buttler, EPFL, Pr Suren Erkman, Université de Lausanne
Dr Théodore Besson, Université de Lausanne
Stéphane Villard, ENSCI-Les Ateliers

COORDINATION
Nicolas Henchoz, EPFL+ECAL Lab

[1] A cross section of the kettle that is bio-inspired by a nautilus, a toucan's beak, and termite mounds. Guillian Graves.

[2] The inner space's fragmentation retains the heat of only the necessary quantity of water. Guillian Graves.

[3] One element of the tea set. The design is not limited to merely translating a scientific result. The result is transformed into an expressive force that gives value to the heated liquid beyond the usual aesthetic codes applied to household items to confer an ecological identity. Guillian Graves.

[4] and [5] In order to answer the project's challenges, the designer creates his own language to express biological mechanisms that have been identified in the living world. He then translates this language into concepts, hand-in-hand with the engineer. Guillian Graves.

The Birthplace of an Adventure

The news came on January 28, 2013: the Human Brain Project is named one of the two flagship European research projects, with a budget of roughly one billion euros. The project combines biology, engineering, information technology, the basic sciences and medicine, to stimulate the human brain in order to better understand how it functions. Top scientists from around the world will be able to test their hypotheses about the brain, the illnesses linked to it and therapeutic strategies. The project originated in EPFL, located on the banks of Lake Léman in Switzerland, near a small city of 130,000 inhabitants. How did this school compete with the best universities on the European continent to succeed in placing two of its projects among the top six finalists? Part of the answer lies in the history of EPFL – a history that is unique in the academic world. Two features in particular stand out: a strategy essentially geared toward transdisciplinarity and exceptional growth over the past decades. Its flexible organization and academic culture seek challenges that bring together many disciplines, providing fertile ground for initiatives requiring multiple skill sets – from basic research to bringing technology to society. Hence, it was an ideal place to revisit the notion of disruptive innovation through the design.

From a Simple School to a Small City

The origins of EPFL go back to 1853. At that time, it was a private institution based on the model of the Ecole Centrale de Paris, which responded to the need to "encourage the development of the practical application of recent major scientific discoveries" voiced by industry. The school opened with eleven students and six professors. Ambitious despite its small size, the school found its way by partnering with the University of Lausanne. More than a century after its creation, in a twist of fate, it regained its independence, becoming the second Ecole polytechnique fédérale in Switzerland. The first (ETH in Zurich) enjoyed this federal status from its founding in 1855, allowing it to assert its global ambition from the start. To date, ETH has seen a total of twenty-one Nobel Prize winners pass through its doors, including Albert Einstein. In the 1960s, the Swiss Confederation recognized the need to develop a similar school in the Latin part of the country. The Ecole polytechnique fédérale de Lausanne took its present name in 1969, with some 1,200 students. This change resulted in spectacular growth: in 2000, the school listed over 5,300 students, including 350 PhD students. In 2013, these figures rose to 9,400 students and 2,000 PhD students. The face of the campus completely changed as well. The site, chosen during the federalization in 1969, consisted of a 52-acre potato field. Today it is a small city of 13,000 that abuts the University of Lausanne (with a campus of 15,000). It also has its own subway stop, a 3,000-seat conference center, a hotel and Innovation Park which is home to more than 120 companies. The EPFL campus has become a hub of knowledge, and is often referred to in national and international development strategies.

Between Sky and Sea

While EFPL teaches classic scientific disciplines in the basic sciences, engineering and architecture, it also stands out for its nose for adventure. At the end of the 1970s, Professor Jean-Daniel Nicoud perfected the first computer mouse device by using a ball system. A graduate of the school, Daniel Borel, turned this technological breakthrough into a worldwide success, founding the company Logitech. In 2008 the company produced its billionth mouse. Continuing in this spirit of adventure and innovation, EPFL advised Bernard Piccard's team which, in 1999, succeeded in making the first non-stop trip around the world in a balloon. The school later partnered with Team Alinghi, created by Ernesto Betarelli, to win the America's Cup, the most prestigious sailing regatta in the world, where giant sailboats – Formula 1 race cars of the sea – compete for the world's oldest sporting trophy. In the 150 years since its inception, no European country had ever won the trophy. Alinghi – with the help of EPFL – won the challenge in its very first participation in the Cup (and even managed to hold on to the title the following year, in 2007). The partnership with L'Hydroptère sought another level of performance. The 18-meter boat equipped with hydrofoils (small wing-like structures that raise the hull up and out of the water) had been trying to beat the world sailing speed record for 15 years. In 2009, with the support of private banker Thierry Lombard and EPFL's help, Hydroptère reached an average of 51.36 knots over a distance of 500 meters, at times with a speed of up to 55 knots – more than 100 kilometers an hour! EPFL was also involved in launching Solar Impulse, the first manned solar airplane capable of staying in the air for several days and nights. While technical performance was essential for this feat, the environment and human beings also played a critical role. These examples illustrate the School's interest not only in scientific results, but also in projects that are part of a broader context.

These challenges confirm the need to go beyond the boundaries of the each discipline's culture – between cutting-edge science, which requires sophisticated calculations, and sailors confronted by the elements, between freedom of academic research constantly seeking the unexpected and the actual object designed to accommodate people on board. EPFL has thus developed a special flair for transdisciplinary action; over the course of its history, it has created new domains that lie at the junction of traditional subjects. This is evidenced by the addition of the microtechnology department in 1978, with a comprehensive field of study and, more recently, with the introduction of life sciences at the crossroad with engineering and basic sciences. With its rapid growth and modest origins, EPFL has remained unburdened by heavy institutional traditions, and is continually reinventing itself around emerging themes – like a Post-Modern society freed of the great founding doctrines, that has made the individual the core of its enterprise.

Architecture of Encounter

In the previous chapter, Michel Maffesoli called Roberto Venturi's architecture the first physical manifestation of postmodernity. Echoing this, EPFL's physical development is manifest by the evolution of its architecture. The Rolex Learning Center (designed by the Japanese agency Sanaa) embodies this evolution, with a global impact. Several months after its inauguration, its designers received the Pritzker Prize. *Wallpaper* magazine named it the best public building of the year. The project's origins date back to the early 2000s. With its rapid growth, the campus needed a new student center that would unite people, regardless of their skills or academic discipline, to further the school's evolution. The library was no longer big enough, and the growth of the different departments and their organization by subject limited exchanges. In 2002 the campus underwent a major reorganization:

departments were grouped into large transdisciplinary schools. One of the schools, for instance, brought together architecture, civil engineering and environmental science. In addition, new centers aimed to bring together laboratories from different schools around major projects (such as Solar Impulse or Alinghi). But in order to live this cultural revolution on a daily basis and respond to new learning needs and the critical lack of space, the school needed a building that could literally become the heart of the campus. The project had ambitious goals, namely assembling the books from all the campus libraries into one place for easy access to all knowledge on campus. The idea was for people to be able to work alone or in groups, eat, have a coffee, buy a book, relax, chat – in other words, spend the entire day there, if they so wished. However, the building – while devoted to the rigors of scientific knowledge – nevertheless needed to break free from the tradition of silent, sterile, individual study spaces, and propose an innovative, open, popular environment conducive to creation, exchange, spontaneous interaction and the development of new communities.

The architectural competition quickly became a major event. After an initial "open" round, twelve firms were selected for the second round, including such respected names as OMA/Rem Koolhaas, Jean Nouvel, Zaha Hadid, Herzog and de Meuron, Diller Scofidio + Renfro. Sanaa (Kazuyo Sejima and Ryue Nishizawa), a Japanese firm, surprised everyone by winning the jury's unanimous vote. Their project was a radical departure from the rationalist approach: the building would be a single level that would rise and fall in different places, undulating above the ground like huge concrete waves up to 90 meters in length. Its interior design far exceeded specifications, proposing large open spaces for relaxing, chatting, etc.

Major private partners supported the project, including Rolex, Logitech and its founder, Daniel Borel, Nestlé, Crédit-Suisse, Novartis and Sicpa. The architects' approach caused debate on the EPFL campus: the design was not the result of a logical, rational analysis of functions, but rather a social vision of interactions and practices – a novel expression of space and a unique relationship between interior and exterior. A single point of entry (located at the center of the building) would encourage informal, impromptu gatherings and encounters. Large arches would make it possible to access the entrance from any direction by passing under the structure. The various features were indeed in keeping with a large, open, fluid space. The vast patios on created by the different "levels" of the slope would create different ambiances – from the tranquility of an individual study area to the buzz and bustle of dining and meeting areas. Free-flowing organic spaces would encourage spontaneous encounters and unexpected interactions that would transcend established communities. Finally, the project would fundamentally challenge the relationship between the building and its environment. Certain areas would allow visitors to recline and interact freely – as one might in a park – and offer individuals a more or less constant link to both nature (notably the mountains) and the imposing campus, with its 300 laboratories. The building, in other words, would both unite differences and provoke contrast. Without deviating from the rigorous specifications, it would bring a new dimension to a school of largely researchers and engineers, who, in their respective disciplines, must meet the rational requirements of scientific construction.

Inaugurated in 2010, the building (now known as the Rolex Learning Center) has had a palpable impact on campus life. It also attracts other great architects for the design of new buildings. In 2012, Dominique Perrault started two new projects. In 2014, Kengo Kuma will take over with "Under one roof," a project with a strong cultural dimension. A unique, vernacularly-inspired roof shelters the Montreux Jazz Café, an exhibition space linked to the Gandur Fondation pour l'Art and a welcome center. This new project will invite many laboratories to contribute to the social dynamics, the development of collaborative works and artistic creation that will bring this new project to life – an ideal testing ground for a laboratory at the crossroads of design and technology like the EPFL+ECAL Lab, whose development will be discussed in the next chapter. Scenographic devices, acoustic technologies and systems of interaction and visualization – the potential for collaboration between research and cultural experience opens up countless opportunities for the campus. The site's architectural and urban pertinence both express and strengthen the school's transdisciplinary nature and its ability to foster the emergence of new areas of study. The Human Brain Project gives it an unparalleled dimension, and the environment itself sets the stage for other projects also at the junction of traditional disciplines, scientific research and innovation.

The Milking Stool Revolution

A few miles from the EPFL campus, another school – the University of Art and Design, Lausanne (ECAL) – also experienced a profound transformation. Artist Pierre Keller, who took over as director in 1995 in a very different context, breathed life into the school in the form of unprecedented growth: the number of students has tripled in fifteen years and now stands at over 600. Here, again, a new building helped incarnate this metamorphosis, this time with the help of another renowned architect, Bernard Tschumi, best known as the designer of the Acropolis Museum (Athens) and the Parc de la Villette (Paris). For this project, he collaborated with Serge Fehlmann. Pierre Keller's instinct demonstrated uncanny intuition. Like the president of the EPFL, his strong Swiss roots combined with an international career make him fearless when it comes to breaking with tradition, freeing him from the great principles of the past.

Il Salone del Mobile in Milan is the biggest annual design event in the world. Visitors to the 2002 Fair recall Pierre Keller and the ECAL's milking stool exhibit at the Swiss Cultural Center. Every designer dreams of creating a chair for a famous brand. And that is just what Peter Keller did – but in his own way. He took as a starting point the "botte-cul," the milking stool so typical of rural Switzerland (the one used by farmers to get all that good milk so essential for cheese and chocolate). This chair is comprised of a circular base on a single leg, with a leather belt which the user attaches for optimal movement from cow to cow. The ECAL asked its students to rethink it in a very open way, and likewise invited a number of renowned designers to participate (Konstantin Grcic, Ross Lovegrove, Jean-Marie Massaud, Yves Béhar, the Campana brothers, the Bouroullec brothers and Hannes Wettstein, to name just a few).

Pierre Keller proudly opened himself to fervent critique, with a chair as black as it was phallic. The designer's cult-object collided with Swiss pop culture to create a surprise sensation and receive unprecedented attention. With a team of young designers from the school, Keller instills his students with the strength to express a radically different outlook on our daily lives, and to question objects and our relationship with them. He avoids mere allusion, to create designs that are both constructive and highly communicative. The ECAL has many projects and exhibitions – from conductors' batons to survival tables to music boxes – all of which create an echo in a media in search of novelty. The School is also cultivating relationships within the private sector, especially the luxury goods industry, thus offering it a great freedom. Partnerships with Boffi, Christofle, Alessi, Baccarat and many other leading brands anchor this approach in the reality of production. Certain designers, spotted by Pierre Keller at the dawn of their careers, regularly return to the ECAL as workshop leaders, despite the demands of celebrity. Hence, in just a few short years, the ECAL has forged both a strong network and a notable reputation. Its appeal allows it to be increasingly selective, with regard to the students it accepts, and to launch new courses of study, particularly at the master's level. In just a decade, industrial and object design has become one of the school's leading departments, along with visual communications. Like the EPFL, the ECAL has gained greater freedom, and while it continues to cultivate basic disciplines like typography and photography, it is not afraid to break from the established order and blaze new trails with innovative projects and new areas of investigation.

Interview with Pierre Keller
Beyond Function

On his patio in St-Saphorin – a small, mythical Swiss village within the UNESCO world heritage-listed vineyard of Lavaux – Pierre Keller evokes with extraordinary verve his career and the development of ECAL (University of Art and Design Lausanne). In fifteen years, ECAL has risen to become a major reference in design training – fifteen years during which Pierre Keller set aside his work as an artist to focus his energy on the University. Too often forgotten, his work as an artist has recently resurfaced, with several exhibitions (most notably at the Espace Lausannois d'Art Contemporain (ELAC) in Switzerland and the Maison Européenne de la Photographie in Paris) and a book. There is, of course, his work as a graphic designer (primarily posters and typographical designs), but there is also and above all his personal work with Polaroids, drawing on their immediate power, intuition and intensity. The provocative themes – for example, his photographic studies of horses' hindquarters – have at times sparked an almost virulent reaction. Though his images may be somewhat destabilizing, they nonetheless open up original perspectives and new ways of seeing, striving to convey an emotion or sensation that is simultaneously powerful, animal and instinctive, that is as subtle as it is unstoppable. Pierre Keller manages move beyond paradoxes. He stimulates his ECAL students with his boundless generosity, but is intransigent when it comes to nonchalance or platitude, taking on endless challenges with exuberance and accepting them with the common sense of one who works the land. In 2011, under State regulations, he handed the reins of ECAL over to designer Alexis Georgacopoulos. In addition to his various exhibition, conference and teaching activities, he is now responsible for promoting Swiss design.

Many of your actions and decisions seem to have been guided by intuition – and not only in the fields that you know best. How do you explain this?

Intuition has guided my entire life – my work, my business and my relationships. There is a notion of common sense and experience. In art and design, it's about training the eye and how we see. What interests me is discovering new images – not necessarily ones I like, but things that have never been done before. Sometimes, I think that someone who manages to irritate me on several occasions must have something strong that I should take an interest in. There is too much repetition, just done differently. You have to be selective.

But how does one learn to be selective?

I don't think I learned much at school. At the time we were trained as artisans. It was when I was traveling that I started to develop my knowledge. In 1965, I was a pretty rational person. I hesitated between Germany and Italy, and eventually opted for Italy.
I was lucky because there was so much going on there. I found myself working on a book with Umberto Eco. When I exhibited at La Polena in Genoa, Angiolo Giuseppe Fronzoni did the catalog. He was one of those people involved in architecture, design and graphics. I was able to rub shoulders with the people who radically changed our vision of design, like Ettore Sottsass and Alessandro Mendini. It was an amazing time where everything was moving fast. In product design, Olivetti played a pivotal role, like Apple today.

You worked in graphics, you're an artist, but at ECAL you've focused on industrial and product design...

When I took up my post at the ECAL, I initially focused on reforming graphic design and the visual arts. Then I realized that we really had to change the approach to design, which was primarily ergonomic. I said to myself that this was going nowhere; we had to start doing things differently, finding ideas that would have an impact without being limited to just function and being able to go as far as making things that were useless. We needed to move away from a purely linear approach to teaching that would surprise the students. A school should be a place of pleasure and research. It needs to be able to provoke in order to remain on the forefront of a discipline's routine.

But how does one avoid slipping into futility and gimmicks and manage to keep a certain perspective?

It's thin line. It's one thing to break with conventions and move beyond what's accepted, but transgression also has to serve the project, it has to ring true and be intelligent. I often say that we shouldn't take ourselves too seriously, but *do* things seriously. This approach quickly convinced young designers with a lot of talent, like the Bouroullec Brothers. They agreed to work with the students.

When you first started teaching design at the l'Ecole Polytechnique Fédérale de Lausanne, did you have to adapt your discourse to motivate young engineers?

We found ourselves in front of an audience of 400 people, with Alexis Georgacopoulos, who was Head of Industrial and Product Design Department at the ECAL at the ECAL. The students seemed bored and were typing away on their computers. We took it for amateurism. It was hard work adapting to a very different culture and context. I finally ended up banging my fist on the table, and made them close their laptops. Then we introduced them to some of design's key players, inviting [the latter] to give conferences. This made what we were saying come to life. Then we started up workshops, with much smaller classes, obviously. This changed everything, and we saw the potential of this kind of interaction.

With the benefit of this experience, how do you see the relationship between designers and engineers?

There's fantastic potential, and what we've done together proves it. Access to new technologies gives us the chance to work outside of conventional themes. But we mustn't be misled: to each his own craft. There is a tendency for everyone to think they know everything. The most dangerous thing is when designers take themselves for engineers, or vice versa. Tools may have become more accessible, but they don't give you the vision, the culture or the trade.

Before working with the EFPL, you'd already confirmed your role in innovation with a well-targeted partnership strategy.

Yes, we did a lot of work with the Italian houses, as they're often still family-run outfits. Their independence gives them the freedom to have a more artistic approach. They still use true skilled craftsmanship. Alexis Georgacopoulos is continuing this strategy with Vacheron Constantin, for example. We must combine our new ideas with this craftsmanship. The same applies to technology.

Project

Give Me More

GIMME
MORE
IS AUGMENTED
THE REALITY
NEXT MEDIUM?

[2]
[3]

How is the price of a luxury purse determined? The design, of course, the material it is made from and manufacturing all define the purse's physical quality. But the principal value lies in the dream the object symbolizes, the story it tells and the status it confers on the person carrying it. This immaterial value is created through powerful communications strategies with VIP events, star-studded advertising campaigns, exhibitions, publications, etc. Instinctively, clients associate the image created with the object they see in the store. But these images and events are still far removed from the client and the purse itself. An object that could directly express its history and virtual qualities in real time, in the client's hands, would radically change our relationship to the physical world. Augmented Reality (AR) techniques allow virtual content to be added to objects, architectural spaces and people in real time. On a very close screen, for example, we can see the object we are holding in our hands enhanced with information or digital animation. In some cases, the animation can even be projected onto the object itself. The basic principle works using a camera that films the object. A computer programmed to recognize particular patterns on the object analyzes the image. As soon as it detects one of these patterns, it adds the prescribed virtual content to the filmed image. Quickly and reliably recognizing a pattern is a formidable scientific challenge. The EPFL's Computer Vision Laboratory, run by Professor Pascal Fua, has played a pioneering role in this field. In the early 2000s, Julien Pilet and Vincent Lepetit established the principles to carry out the operation without having to overlay the object with markers – the famous "QR codes," so easily identifiable by their arrangements of black and white squares. Thus, for the user, the objects remain intact and there are no longer any artificial signs to disrupt the link between the physical and virtual worlds. The promises of AR have led to a great many demonstrations and filled many a scientific conference hall with – for the best among them – results that seem like magic. AR has become a part of everyday life, whether for a playful effect on chocolate boxes or to add an extra dimension to game consoles such as the Nintendo 3DS. [4] →

However, one question was rarely addressed during this period: how to bump this technical prowess up to the rank of a veritable form of media. In its commonly-accepted definition, a form of media only exists if its narrative force – its capacity to express ideas and

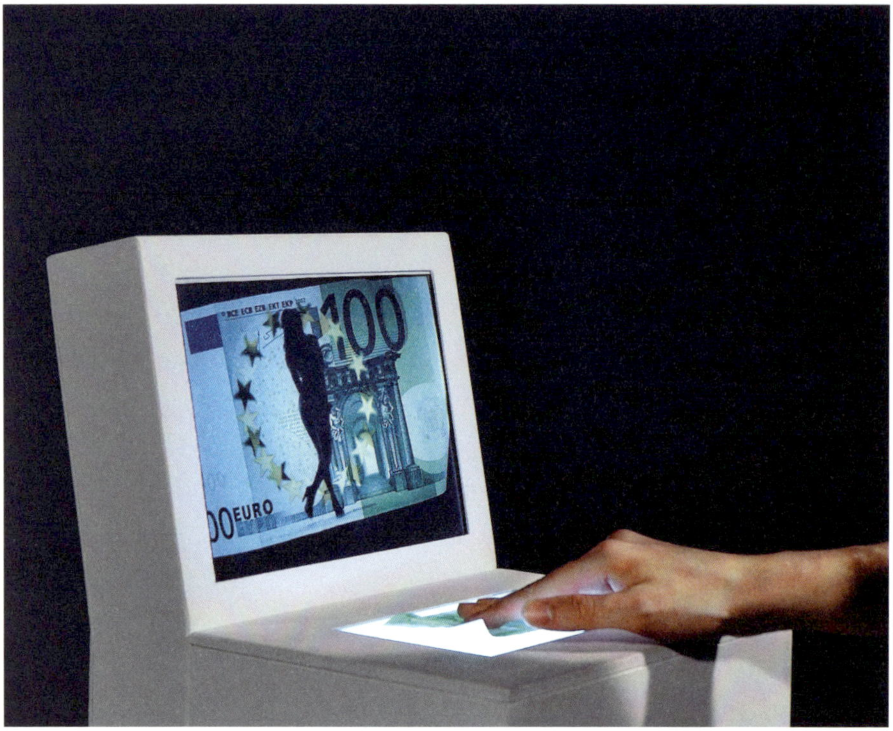

[5]

values – takes precedence over the technological display. In 2007, Camille Scherrer, an ECAL student, worked with Julien Pilet for her diploma project with backing from the l'EPFL+ECAL Lab. In 2008, her book, *Le Monde des Montagnes*, won the Pierre Bergé Prize for the best European Design Diploma. A great art collector, Yves Saint Laurent's companion is not an IT or digital performance enthusiast. However, he was won over by the idea and the story, which was more important than any technical flamboyance. Part of Camille Scherrer's family originates from the Swiss mountains, more specifically the "Pays d'Enhaut" or Highlands. She compiled a collection of family photos taken at the beginning of the 20th century to create a very classic family album. Placed on a sober table equipped, naturally, with a desk lamp and computer, the images suddenly unveil a hidden meaning and reveal an extra dimension expressing the author's imagination. She did not rely on spectacular 3D effects, focusing instead on the evocative quality of her work and the relationship between the book's pages and the animation. Her animations were based on a succession of cut-out paper images. The relationship with the book's material became legitimate, and there was also a cultural justification: paper cut-outs are one of the Swiss Highland's traditional arts.

When she had completed her diploma, the EPFL+ECAL Lab hired Camille Scherrer to formalize the partnership and explore in a more systematic way the narrative principles of AR, its forms of expression and its visual languages. Many young designers contributed to the project, which also attempted to comprehend different use contexts – from the object to the architectural space. The Laboratory has brought together the most convincing experimental installations in a constantly-evolving permanent exhibition entitled Give Me More. In 2010, the exhibition was awarded best prize at the International Design Festival in Berlin. The interesting thing was that the jury did not include a single interaction design specialist. As with the Pierre Bergé Prize, it was the stories themselves, the quality of the interactions and the experience offered to the user that won over the jury. AR was beginning to play its role of a real media form. Research intensified at the laboratory with new designers who explored the relationship to the body and the notion of identity. The Tattooar installation encourages visitors to get a temporary transfer tattoo on their arm. In front of a digital mirror, the tattoo begins to change, becoming animated.

The project calls into question not only the virtual value of the object but also the notion of its permanence. Like a jewel, a tattoo does not change over time. Can they be enhanced, given an evolving dimension thanks to the virtual world's dramatic arrival on the scene? Tattooar also explores the idea of immersion. Most brands want to put the user (in other words, the client) at the center of the experience, but they associate a digital installation's power of immersion with the size of the screen. The EPFL+ECAL Lab's work shows that other principles offer equally significant immersive potential: animation that seems to emerge from our flesh draws the user in, as does calling upon the user's imagination to create a story. Liron Kroll explores this aspect with her project Last Year. She proposes a non-linear narrative principle; rather than create one image that leads to a complete virtual story, she arranges a series of objects. With the help of a tablet, users roll over the object in the order they wish. Each object causes a virtual sign to appear in the form of a short animation. From these fragments users naturally imagine an entire story, closely related to their personality and their experiences.

In 2011, Give Me More gave rise to an initial summary at the International Symposium for Mixed and Augmented Reality. The contribution, acclaimed by the event's jury, presented a set of principles (narrative principles, interaction strategy and visual language) that would help turn AR into media. Research continues today, with concrete applications such as the Seeds of Colors installation for the Brazilian group Melissa, described in this book. In January 2013, the renowned Eyebeam Gallery in New York's new district of galleries next to the Gagosian and Gladstone, devoted its 500 square meters to Give Me More, since renamed "Gimme More" to give it a local feel. Under Michael Szivos's guidance, the SoftLab team brought a new vision using an original stenographic approach, with the collaboration of Pentagram. Curator and journalist Laetitia Wolff used Gimme More to launch a major debate in the gallery about "Is Augmented Reality the Next Medium," with important figures from the world of art, design and innovation.

[6] These results are born of the encounter between the new computer [7] → vision technologies developed by science labs and the exploration through design of their potential uses, their contextualization and the resulting principles of expression. They go beyond "magic" to provide the basis of a new medium that is capable of

AQUARIAN

enhancing content. This encounter is broadening the issues, and the prospects remain considerable – for the luxury goods industry, for example, which devotes colossal resources to developing the virtual value of its products, and for other products that distinguish themselves by their environmental and societal qualities. AR offers a new avenue of communication; directly associated with a product, it will be able to stand out from the abundant writing and symbols on packaging. It also provides new areas of artistic expression, whether through enhanced books, performances or installations. Many trials are held in museums to attempt to change the relationship between the visitor and works of art. It also has considerable potential on a wider scale, for our relationship with living areas and urban information. Overall, AR will profoundly change our way of relating to the physical world, enhancing information by associating it with a new capacity of expression. However, we must continue the work to give this media a specific language that allows it to express relevant stories and practices that give it legitimacy in our everyday lives.

PROJECT
Nicolas Henchoz, EPFL+ECAL Lab

INTERACTION DESIGNERS
Cem Sever
Vincent Jacquier
Camille Scherrer
Thomas Eberwein
Liron Kroll
Yuri Suzuki
Happy Pets
Marc Mussler
Thibault Brevet
Yuri Suzuki
Maria Laura Méndez Martén
Angelo Benedetto

INDUSTRIAL DESIGNERS
Susanne Schneider, Nicolas Lemoigne,
Béatrice Durandard

PROJECT ENGINEERS
Daniel Tamburrino, Olivier Nguyen,
Gavrilo Bozovic, Fanny Riedo

SCIENTIFIC PARTNER
Computer Vision Laboratory, EPFL, Pr Pascal Fua

RESEARCH ENGINEERS
Julien Pilet, Vincent Lepetit

[1] Renamed Gimme More for New York, Michael Szivos and the Softlab team's scenography imparted an elegant staging for the project in the vast space at Eyebeam. Photo Allan Tansey.

[2] DOTS
Maria Laura Méndez Martén.
Childhood memories – remember linking a series of dots following their numbers? Here the adventure goes further as the drawing then comes alive. The boxer starts fighting and an uppercut will knock out all the numbers which served to create him! Here the user takes an active role by creating the pattern to be augmented. Photo Christophe Fillioux.

[3] STITCHED PIXEL BY PIXEL
Camille Scherrer
Augmented Reality permeates dreams emerging from a cushion. Here the visual language of embroidery, the cross-stitch, meets that of the digital world, the pixel. Thanks to the research of Julien Pilet, AR adjusts to deformable surfaces and places the animations in the users' hands for even greater realism. Photo Sophie Brasey.

[4] GHOST
Thomas Eberwein, Tim Gfrerer,
Music by Simon Pyke
A snow storm is raging within an abandoned, barren landscape. Inside this storm the viewer can make out a procession of human bodies which seemingly try to find their way out. These bodies are the remains of previous viewers of the installation, their ghosts, forever trapped in this hostile environment. *Ghost* emerged from an extensive research into the interaction between the user and a virtual environment.

[5] CASHBACK
Vincent Jacquier
ECAL
In *Cashback*, visitors simply place a European banknote of any denomination on the installation's glass to see the figures pictured on the note come to life in a surprisingly provocative way. The higher the value of the note, the greater the payoff in terms of content yielded. *Cashback* shows how one can alter the perception of an object by allowing it to express virtual content directly linked with its physical presence. Photo ECAL.

[6] TATTOOAR
Cem Sever, Happy Pets,
Marc Mussler, Thibault Brevet
Cem Sever invited three tattooists reputed for their graphic talent to create tattoos that come alive in front of a digital mirror. This installation questions identity through the symbolic strength of the tattoo – an image digitally inlaid in the flesh of the user. The tattoo takes on a life of its own and its partly virtual nature enables it to evolve over time. The relationship between the durability of the physical object and a possible evolution of its expression, content, and meaning is a key topic that is even applied to our bodies. Photo Christophe Fillioux.

[7] BEATVOX
Yuri Suzuki
Yuri Suzuki has created an installation that harnesses the principles of augmented reality to interpret them in the realm of sound by enabling users to control a set of drums using only their voices. Unlike the exhibitions' other installations, the augmentation in Beatvox is physical rather than digital. Photo Michel Bonvin.

Design and Technology
The Encounter

At the turn of the millennium, while EPFL and ECAL were both undergoing a metamorphosis that brought classic disciplines face to face with new horizons, the notion of transdisciplinarity occupied many minds, fueling many discussions and events. The International CREANDO Symposium (Creativity in Economics, Arts and Science) in Zermatt embodied this euphoria, enhanced by the thrill of mountain peaks and champagne bubbles. Created in the 1980s, this symposium attracts major players from the worlds of science, art and management in a spirit of communication that overcomes cultural and geographical boundaries – a kind of globalization of the mind.

Rare are those who dare question the relevance of certain interactions. However, in 1993 – its 7th year – mathematician Mitchell Feigenbaum (known for the universal constant for functions approaching chaos, to which he gave his name) explained how chaos theory can predict the behavior of many systems, including train networks, the weather – even the number of fish in an aquarium. More specifically, he explained how it can be used to determine whether, in the event of a disturbance, these systems will behave in a stable (and thus predictable) or unstable (and therefore chaotic) manner. At the end of his lecture, during the discussion period, Feigenbaum broke taboo when a well-known jazz musician took the words Feigenbaum had used during the lecture to draw a direct parallel with God and death. Not yielding to the temptation of allegory, Feigenbaum responded by saying that he had merely presented a mathematical theory – the result of a scientific approach; hence, while he would leave his interlocutor the freedom of his amalgam, that he himself saw no real connection, thus abruptly ending the discussion. The audience was stupefied. Why had Feigenbaum compromised this magical atmosphere of intellectual communion? The moderator for the event (a psychologist by profession) did his best to re-enliven the spirit that had reigned prior to the incident.

This anecdote raises fundamental questions, the likes of which preceded EPFL and ECAL's collaboration: how to implement a transdisciplinary initiative that reflects the differences of language, culture, reference and temporality of each discipline? How to avoid the lowest common denominator syndrome, which reduces each participant's contribution to what everyone already knows from the start? The challenge is even greater in the academic setting, as academic culture does not aim to develop solutions that meet a particular set of specifications. Researchers initiate projects based on solid, sound assumptions that nonetheless are intended to capture the unexpected and lead to the emergence of original knowledge. With such openness and freedom, managing projects and interactions becomes much more challenging. The question of time illustrates the type of misunderstanding that can arise from such a meeting. For an engineer, academic research is more often than not based on a period of four years (the duration of a typical doctoral thesis). The process is long, rigorous and has many phases (a well-researched bibliography, a theoretical hypothesis, an experimental principle, the building of a measurement tool, experimentation, analysis, comparison with a theoretical approach, publication and verification). This process – completely legitimate in the scientific world – seems slow to designers; for them, this quasi-asocial behavior is in no way synonymous with creativity. They do not see the intensity of their own approach – often based on inspiration, a set of visual references (and their reappropriation), the feel of a certain material or form, the simplicity of function, or concepts borrowed from other disciplines –, especially when partaking in a project that openly explores, and is not dictated by the demands of a specific industrial project.

Conversely, the designer's approach breeds mistrust in scientists, who above all see a superficial, cosmetic, ephemeral, aesthetic method that plays on effects of style, thus illustrating one of the many possible misunderstandings between design and engineering (and hence, the temptation to compromise, which poses a threat to the contributions of both disciplines). Coming back to the concept of time, a timeframe that, in the scientific world, would result in shoddy or unfinished work, would dull the designer's expressive strength by diluting his or her efforts. The meeting between EPFL and ECAL, both of which have enjoyed tremendous freedom of action and remarkable development, should forge a new model of collaboration – one that is capable of uniting these disciplines while leaving room for the other to express its own uniqueness. This also means leaving enough time (again!) for everyone to do their job, but targeting periods of intervention and interaction. While this seems both obvious and legitimate, its actual implementation is another matter altogether, and requires many stages and much testing and experimentation.

The First Challenge

The first meeting between the EPFL and ECAL took place in 2001. Nicolas Henchoz, then director of communications for the EPFL, called upon David Rust and Gilles Gavillet (former ECAL students) to rethink the EPFL's visual identity. The ideas they brought to the first publication were radically different; their choice to use images inspired by neo-Realism broke with artificial representations to reveal the university campus from the inside, with details epitomizing laboratory life. In the text sections, Rust and Gavillet played with a new typography (handwritten titles and rapid highlighting effects – evoking student life – and radical variations in paper types). The reception held for the release of this provocative publication brought together the heads of the two institutions, Pierre Keller and Patrick Aebischer (who immediately hit it off) for the first time, giving birth to the first collaborative projects. Patrick Keller and Christophe Guignard – two teachers at the ECAL and former EPFL students – addressed the concepts of digital interfaces and the relationship between information, robotics and built space. A series of short workshops allowed the two institutions to launch the first explorations. In 2003, design became an elective for EPFL students, as part of the new program requiring mandatory courses outside of the hard sciences. Among the 23 branches to choose from, design immediately became the students' second choice (after management). Quite a success! More than 400 students enrolled in the Intro to Design course, taught by Pierre Keller and Alexis Georgacopoulos. These experiences fed the relationship between the two institutions.

The search for a new building for ECAL hastened the prospect of a partnership. With its campus spread out over two sites (one in the town center and the other on the outskirts), in rooms overly crowded due to growing number of students and projects, ECAL had no choice but to move. Pierre Keller set his sights on the old IRIL factory, which once manufactured women's silk stockings, and employed nearly 1,500 workers. Gradually becoming unsuited to production costs in Switzerland, only a few offices remained on the site. That (and a few defunct relics of machinery in the vast halls) was all that was left to attest to the factory's former activity. The site had huge potential, but far exceeded ECAL's needs. This, as it turned out, was actually a catalyst, as four institutions ultimately decided to share the space. These institutions include ECAL itself,

ELAC (Lausanne Area Contemporary Art), an international art gallery (created by Pierre Keller) in the form of an association, an incubator created by the town of Renens, which offers spaces to young designers and entrepreneurs to launch their businesses, and one more: Patrick Aebischer, President of the EPFL, also decided to sign his school on for the adventure, with the idea of creating a laboratory to bring the collaboration between the institutions to fruition.

The Birth of the Lab

Nicolas Henchoz presented the idea for this collaboration to EPFL management on March 5, 2007. To build on the existing names rather than creating a new one, the new EPFL unit would be based in the future ECAL building, and be called the "EPFL+ECAL Lab." Its mission would draw on the experiences of its predecessors (the MIT Media Lab and d.school at Stanford, for instance) to propose a new model. The aim was not to create a chimera of scientific research and design but rather a design lab that would explore opportunities and issues related to emerging technologies. Switzerland invests heavily in research; its laboratories are teeming with radically new ideas that generate what is known as disruptive technology (technologies that, by definition, break with known concepts and that strive not only to improve what already exists, but also offer a novel approach). But to what end? Often, such innovation provides new insights to a broad audience. But how to explore these perspectives and, more specifically, the uses to which they may give rise? How to combine technological performance with content, practices and contexts in order to make it meaningful? How to develop its capacity for expression among a target audience?

The EPFL+ECAL Lab's goal is to provide answers to these questions through design. Design's mission goes well beyond the pure aesthetics to which popular culture too often attempts to confine it. Its role here, however, is to explore new uses, put them in the social and cultural context of users and propose creative principles that are instrumental in creating new products, values and services. In order to avoid the pitfalls of transdisciplinarity mentioned at the beginning of this chapter, the new unit seeks to develop projects based on targeted interactions that allow each discipline to express its unique qualities at the highest level, attempting not to dramatize technological performance but rather to make this performance an experience for the user. As such, its proposals tend to be uniquely those of the Lab (i.e., the idea of to move towards an identity inspired by the "Super normal" of designers Jasper Morrison and Naoto Fukasawa). To be part of the user's context, innovation linked to disruptive technology must offer users new experiences that are both meaningful and in keeping with their values and references, so that they seem familiar, intuitive, normal...natural!

From Ideas to Reality

Works at the future ECAL site are advancing quickly under pressure from Pierre Keller and Pierre Nussbaumer, owner of the former IRIL factory. Architects Bernhard Tschumi and Serge Fehlmann are giving the factory a complete makeover, enhancing certain key elements of its structure and industrial past, and clarifying the major thoroughfares. They are also creating openings in order to bring in light, thus moving away from the feeling of floors stacked one on top of the other towards one of architectural unity. The austere facade of the original building has been enhanced with undulations and primary colors – a change that is particularly notable on the façade of the ECAL entrance, which is entirely covered by a metallic lattice structure (like the pattern of the silk stockings once produced at the factory – an implicit tribute). But we could also read it as an echo of the famous Superonda sofa, the icon of a revolution whose undulation literally bisected the block that was modernism. The aesthetic hijacking of this formerly industrial building again questions the rationality of the original object, except that, in this case, it does not undermine the rationality of the original object, but rather adds an extra dimension to it. The architects have transcended this rationality through gesture and form. They have also chosen to preserve some of the original details that attest to factory life, such as the imprints left by the machines on the floor. The transformation of the site thus creates an ideal place for a project whose mission is to bring the rational nature of emerging technologies face to face with the cultural and social aspects of their use. During the first days of October, 2007 (the start of the academic year), the inauguration of the building attracted quite a crowd. In the EPFL+ECAL Lab's part of the building, an exhibition shows evidence of the collaborative projects carried out in previous years. Let the great work begin!

Interview with Patrick Aebischer
Opening Up New Fields

In Patrick Aebischer's office hangs an impressive work of marquetry that evokes DNA – one of the rare indications of the President's personal world. The work is by the artist Yoki, who is known for his stained glass windows, some of which adorn the dome of the Basilica of Nazareth's. Yoki is none other than the father of this doctor and specialist in neurobiology, who became president of the Ecole polytechnique fédérale de Lausanne in 2000. One of his first tasks was to create a new research department and curriculum at the interface between the life sciences and engineering which he then turned it into a resolutely transdisciplinary strategy for the entire institution. His interest in the diversity of knowledge goes far beyond his professional life; it is part of his own experience that began with a family life that saw a parade of authors and artists pass through – a life in which passion for expression in all its forms reigns supreme. Patrick Aebischer chose a scientific career rather than artistic one, but has kept the family sensitivity and passion for the visual arts and music.

When you became head of EPFL, the word "transdisciplinarity" was already very much in fashion, but was rarely practiced in such an ambitious context. What did you hope to bring to it that was new?

What's fundamentally changed is the capacity to work between two different fields. The aim is to give technical tools to people from very specific disciplines – in this case the life sciences. The main focus remains the life sciences, but with specific training at the bachelor level you can access various tools and thus open up new fields of investigation, especially with quantitative aspects.

Isn't there a risk of creating a "jack of all trades, master of none?"

That's obviously the big risk. Transdisciplinarity shouldn't be anything and everything. I have serious doubts about that kind of intermediateness, where you're in no one specific discipline. Being in a discipline but with new tools – that's different.

With work between engineers and biologists, the work remains scientific, and – consequently – in a rational culture with a shared logic and similar approaches. With the applied arts or human sciences, one leaves this proximity behind...

It's true, within the sciences, the thought structures are the same, and skills are easier to acquire. Interfacing with a discipline that calls upon cultural values and emotions is more complex. I think that the notion of a relationship based on tools remains valid, but the artists or sociologists who use them have to be able to reinterpret them and really make them part of their own discipline. Scientists also have to be able to engage in dialogue, but without venturing outside their field of competence. Interfaces like these require more effort, but are going to be the main source of new understanding and innovation. The humanities can, for example, benefit hugely from quantitative tools or emerging semantics analysis techniques. Architecture is a good example, where technology plays an important role in widening the scope of possibilities. The same goes for design. Most notably, new technologies offer major opportunities in our relationships with the physical and the virtual.

The relationship between the physical and virtual worlds is a central theme for the EPFL+ECAL Lab. Do you feel that we're heading towards a dematerialization of the context of our lives?

We'll find a balance between the virtual and physical worlds. Taking a step back, we could say that the content of a book is already virtual in a way, or at least an abstraction. There will be a change, but not a return backwards. We mustn't forget that our brain adapts to new conditions. This plasticity is a fundamental discovery in neuroscience. The key issue lies in how our brain unites the digital and the physical.

So, to summarize, as head of EPFL, your strategic thrust is based on a rational approach, but also relies heavily on intuition.

Of course. In fact, it's crucial. In the end, major decisions require us to feel things.

How can this intuition be combined with rigor and coherence?

First of all, there's experience – the knowledge that makes up an adapted terrain. You must have the courage to use this intuition and, of course, the right context in which to do so. Then you have to test it in the actual context, be able to react quickly and surround oneself with the right people. The notion of time plays a key role – being neither too early nor too late.

Is this a convincing approach in an environment characterized by a culture that is rational first and foremost?

Everyone is sensitive to some degree, including scientists. Beyond that, it's a question of leadership. I think the hardest part is convincing oneself without being mistaken. You also need to have the courage to take the leap and the charisma and tenacity to convince others.

Project

Hidden Carbon

[1]

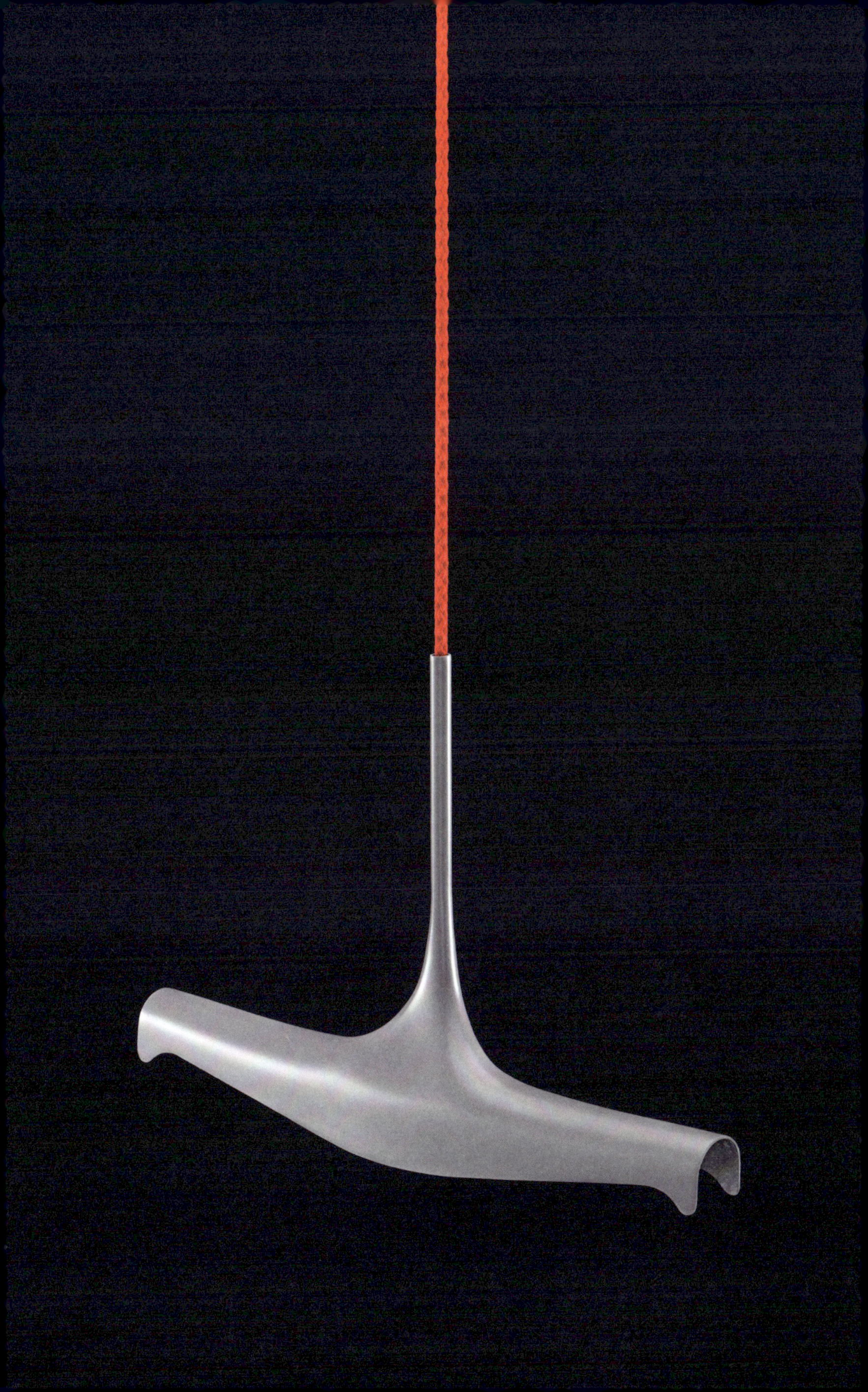

With its pure black shiny fibers, carbon catches the eye. Suspended in clear resin, it resembles contrived fabric (think Chevron), the likes of which one might find in a Haute Couture fashion show. Certain objects made of this material spontaneously come to mind: the Luis & Clark violin (2009) and the interiors of sports cars, to name only two. Hermès even made a saddle out of it. All of which goes to show that it is defined by its properties, carbon inherently embodies performance and high technology. Nearly as fragile as glass if not bound with another material, what it evokes far exceeds its mechanical characteristics. Expensive to make and even more expensive to use, its use is almost exclusively geared towards competition and speed, thus evoking those uses in which it excels (Formula 1 race cars, racing boats and aviation). As such, designers do not hesitate to commandeer it for counter-purposes, replicate its look using *trompe l'œil* or simply boast of its presence. This kind of celebration of technology is reminiscent of the Streamline Style so popular in the United States in the 1930s, where almost any object was given an aerodynamic shape – even household objects – to celebrate the promise of a technological world.

Composite materials with a carbon fiber base indeed have exceptional characteristics, combining lightness and rigidity with form in a unique way. Used wisely, these materials can effectively aid in the quest for performance. For a car, for instance, its lightness allows it to accelerate faster and reduce fuel consumption. This lightness also helps with handling (in tennis rackets, for example). Excluding areas where technical criteria are the primary measure of success, carbon responds above all to marketing goals, which can confer a given object with a unique identity or status, and more importantly generate higher profits.

The performance-oriented, innovative and technological quality of this material poses an especially interesting challenge: can we find new prospects for the use and expression of carbon that transcend this association with high technology? The question was all the more interesting given that the EPFL's Laboratory of Composite and Polymer Technology and Decision Shipyard, founded by
[2] Bertrand Cardis, a graduate of the school, contributed to the many of the challenges discussed in the chapter "The birth of adventure" like Alinghi, Hydroptère and Solar Impulse.

How could this expertise be used to bring new experiences to our daily lives? The EPFL+ECAL Lab launched the "Hidden Carbon"

project with the goal of not showing the material as an end result, but rather to explore its potential in other contexts. For a whole semester an ECAL class (under the direction of Alexis Georgacopoulos and Tomas Krall) and another class from ENSCI-Les Ateliers in Paris (led by Jean- François Dingjian and Eloi Chafaï) tackled this theme which, as it turns out, was less straightforward than expected.

The young design students involved needed to gain an understanding of the material to be able to propose meaningful use for it. But above all the project aimed to make carbon usable in everyday life and offer users new, legitimate, sensitive experiences. The initial theoretical research, typological references and popular inspiration resulted in a first set of intentions. However, in the second phase, a combination of the complexity of the material and its implementation provided another perspective on this potential. Armed with cans of resin and various fiber samples (based on the specifications of their original criteria), the students were able to measure not only the difficulties associated with carbon, but also observe its properties, behaviors and particularities. For instance, in the anisotropy process, the material can become quite rigid in one direction and much more flexible in the other, depending on how the carbon fibers are arranged. But it is essential for designers to observe and feel such a property in order to free the creative process from pure theory, thereby allowing them to balance learning with inspiration and instinct-based approaches. This freedom of exploration was then be confronted by the demand for relevance, most notably by Bertrand Cardis, the owner of Décision Shipyard, the EPFL+ECAL Lab and, of course, the professors involved. No question of producing a simple carbon bookshelf that, once painted, would look no different than the same object made out of steel – just a lot more expensive. For the same volume of material, carbon is not more rigid than steel but much lighter. Visually speaking, its thickness and appearance are almost identical.

Following these discussions, seven projects were chosen for the next stage: the creation of a professional prototype, transforming the ideas and trials into concrete results. The selected proposals
[3] were surprising: Claire Baudrimont's project, for instance, actually contained no carbon in the end – just glass vases with a classic, symmetrical appearance at first glance, but whose shapes were surprisingly flexible and organic, enriched with different textures, both rough and refined. First, Baudrimont cut and sewed carbon

fiber fabric to create flexible molds that were then held in a steel frame. The fiber retained its properties up to a temperature of 1300 degrees during the fabrication of the vases, leaving a fine imprint of the fabric's texture on the glass without adhering to it. Hugo Bartoli revisited the construction principles of bamboo structures. Playing with carbon fasteners soaked in resin to make very rigid bindings, he was able to reduce the amount of material needed to stabilize the structure by half. The result was a vernacular that was both archaic and contemporary, due to the lightness of the visual expression and the savings in terms of material. Quentin Caille revisited the chair – a stylistic exercise so dear to designers. Because of the material's properties, the Quentin chair (borrowing the concept of a swing that can be hung from any beam, tree or urban structure) weighs only a few ounces.

With its wide range of offerings, Hidden Carbon opened up opportunities in many areas, from lamps, to stereo amplifiers to tables, by combining the material's natural properties with proposed uses and forms offering users a unique experience. Following an initial public presentation, Chantal Hamaide (creator and chief editor of the magazine *Intramuros*) invited Hidden Carbon as one of the two exhibitions at the Maison & Objet Fair in Paris in 2011. Carbon had finally left the world of Formula 1 and high tech to become part of our everyday space.

CURATOR
Nicolas Henchoz, EPFL+ECAL Lab

SCHOOLS AND TEACHERS
ECAL/University of Art and Design, Lausanne:
Alexis Georgacopoulos and Tomas Kral
ENSCI-Les Ateliers:
Jean-François Dingjian and Eloi Chafaï.

SCIENTIFIC PARTNER
Laboratoiy of Composite and Polymer Technology (EPFL)

TECHNOLOGY PARTNERS
Décision SA
Suter Swiss-Composite Group

[4]

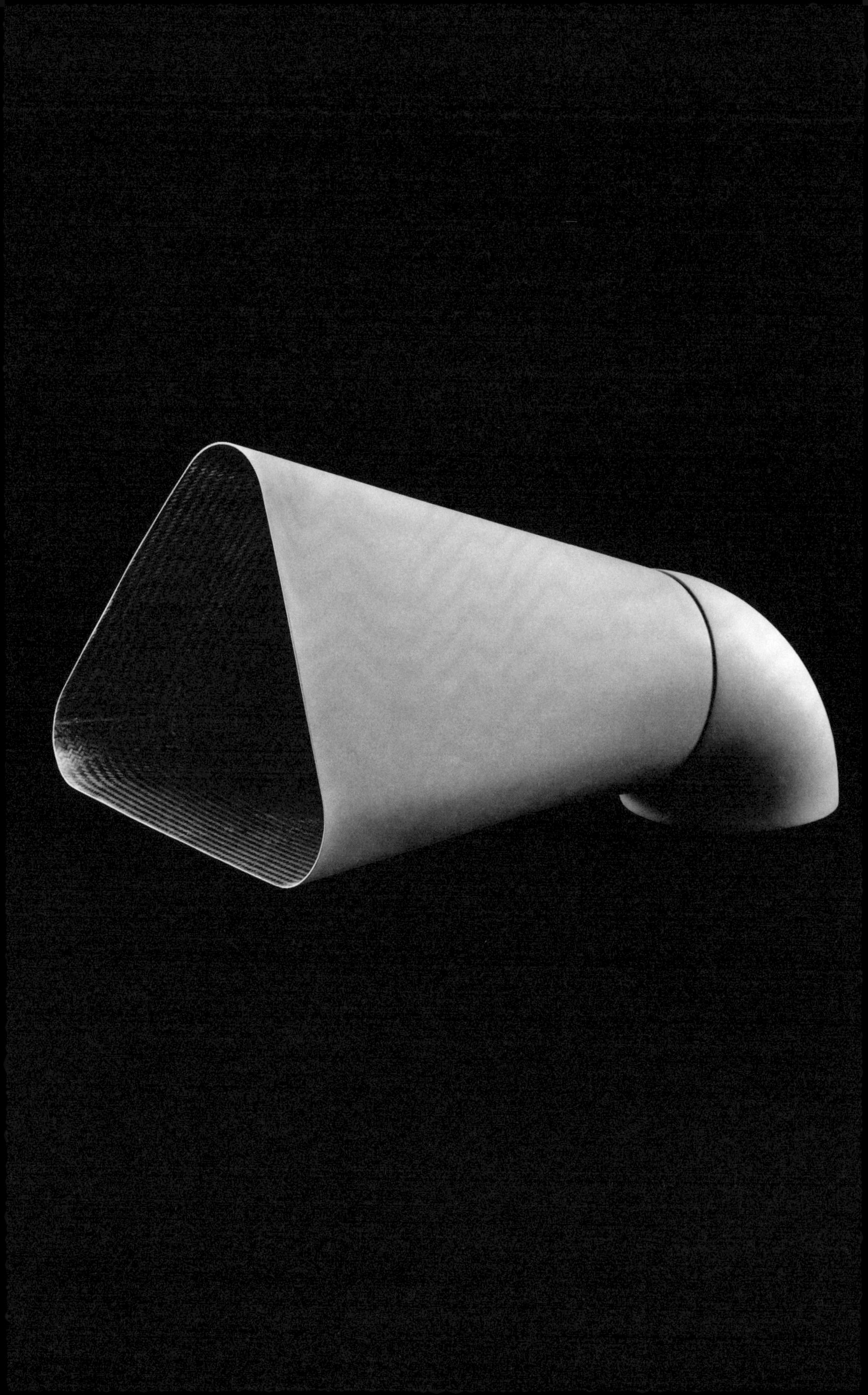

[1] POCKET SWING
Quentin Caille
ENSCI
Weighing only a few ounces, this portable swing turns any tree or beam into a seat. Just passing through? Enjoy temporary comfort almost anywhere, including urban spaces.

[2] TRANSFERS
Claire Baudrimont
ECAL
Glassware blown in molds made of carbon fiber cloth. The imprints that remain showcase a rare alliance between the cloth and fire, without burning.

[3] TRANSFERS
Detail

[4] LINKS
Hugo Bartoli
ECAL
This project combines a natural compound – bamboo – with a technological composite: carbon. Bamboo, a readily available material, is used for the main structure, while the carbon's rigidity allows for light, efficient assembly.

[5] LE CHANT DU CARBONE
Pierre Bayol
ENSCI
Simply connect your MP3 player and the sound becomes magically amplified – without any additional power source! A vibrating object attached to the base of the resonator activates the carbon walls.

[5]

London, Paris, San Francisco

Innovation has no borders. This being the case, the objectives of Lausanne's new laboratory call for a comprehensive approach and a wide range of perspectives. But how to find the creative force and skills necessary quickly and effectively? How to remain flexible in order to address each project and theme in an appropriate manner? The Lab's first major project, Sunny Memories, allowed it to build partnerships with other schools, including ENSCI-Les Ateliers (Paris), the Royal College of Art (London) and the California College of the Arts (San Francisco). The involvement of roughly 85 students and 10 professors meant not only an astounding *number* of ideas, but also an amazing *range* of viewpoints, thanks to the many different nationalities involved in the project and the diverse approaches of schools. Each school has its own specific culture, organization, and methodology fueled by the distinguished individuals that teach in them.

ECAL stands out for the communicative strength of its projects. The painstaking attention to every detail of the objects created conveys the project's intention. Playful ideas, quirky proposals and provocative reappropriation break with accepted frameworks and common preconceptions, making for highly original objects and projects that, nonetheless, have strong cultural roots and that (as regards form and function) are extremely detail-oriented, and achieve remarkable coherency.

ENSCI-Les Ateliers unites students from different disciplines around a common project and greatly structures the initial research work: new challenges lead to multifaceted approaches. Students are organized into groups to discuss the project's goals, define the use context and consider its potential prospects. In the first phase, the class determined the creative field, with a broad overview of the project in light of the tools, materials, and references available. The Royal College of Art uses a more individual approach that encourages radical thinking. It likewise attracts students from a variety of nationalities, most notably Asian. This diversity has an important impact on classroom dynamics. The Royal College also has postgraduate research centers (for instance, the Helen Hamlyn Centre) that use design to address societal challenges through ethnology and anthropology-based research approaches. Such approaches are useful for developing empathetic solutions that take users' perception and needs into account, which is particularly important given the Centre's focus on the elderly and disabled, patient health, work and the city. Its solutions likewise have an inclusive mission – to bring people together, rather than stigmatize the precarious situation of a minority. Such approaches provide an essential contribution to the EPFL+ECAL Lab, which has developed a special partnership with the Helen Hamlyn Centre. The project "You, Me & Everyone We Know" questions the interfaces of social networks relative to the elderly. Through field surveys conducted by Peter Ziegler, an assistant researcher at the Centre, the project has led to important findings on differences in perception regarding digital technology depending on the user's age.

From One to a Hundred Designers

Like "You, Me & Everyone We Know," exploring the prospects of emerging technologies and the issues they raise sometimes requires extensive fieldwork and, in certain cases, is extremely complex. Take, for instance, augmented reality. Adding virtual information to physical objects in real time involves designing the object itself, creating the interactions, digital animation and overall system, not to mention mastery of a wide variety of tools. In other cases, as in the case of "Under Pressure" (a project that centers around densified wood), monumental efforts are needed to turn ideas into reality. First, high quality molds must be made. Then, a number of trials are needed to produce a successful piece (often on lab machines not necessarily suited to production). Projects like "Sunny Memories," however, are more accessible. While the new generation of dye-based solar cells involves highly complex scientific techniques, the challenge for designers nonetheless lies in the elements that form the basis of their visual vernacular: form, color and pattern. These varied approaches and the number of potential projects are key factors of success.

This diversity prompted the EPFL+ECAL Lab to create a specific strategy for each project, determining on a case by case basis whether to develop the project within its team, in partnership with a design school, or with the help of prominent designers (based on mutual interest in new fields of experimentation). The Lab team defines the framework for the project, prepares the material necessary for its realization, and follows the progress of the research and creative work. It also oversees the optimization of the findings (user trials, exhibition tours, documentary work, turning the experience into training modules, issues of intellectual property, etc.). This aspect (which will be discussed in the next chapter) also characterizes the EPFL+ECAL Lab. Since its conception, the Lab has had the distinction of being under the jurisdiction of the Vice Presidency for Innovation and Technology Transfer from the EPFL, rather than of a specific school. Thus, not only does it preserve its freedom in terms of the themes it chooses, but it is also involved in the innovation process – at the interface between the academic world and society at large. In this respect, the scope and diversity of this international network is a key factor for approaching each project with the utmost relevance – spreading transdisciplinarity, developing knowledge, and combining the skills created in technology laboratories with popular culture.

Project

The Colors of Dignity

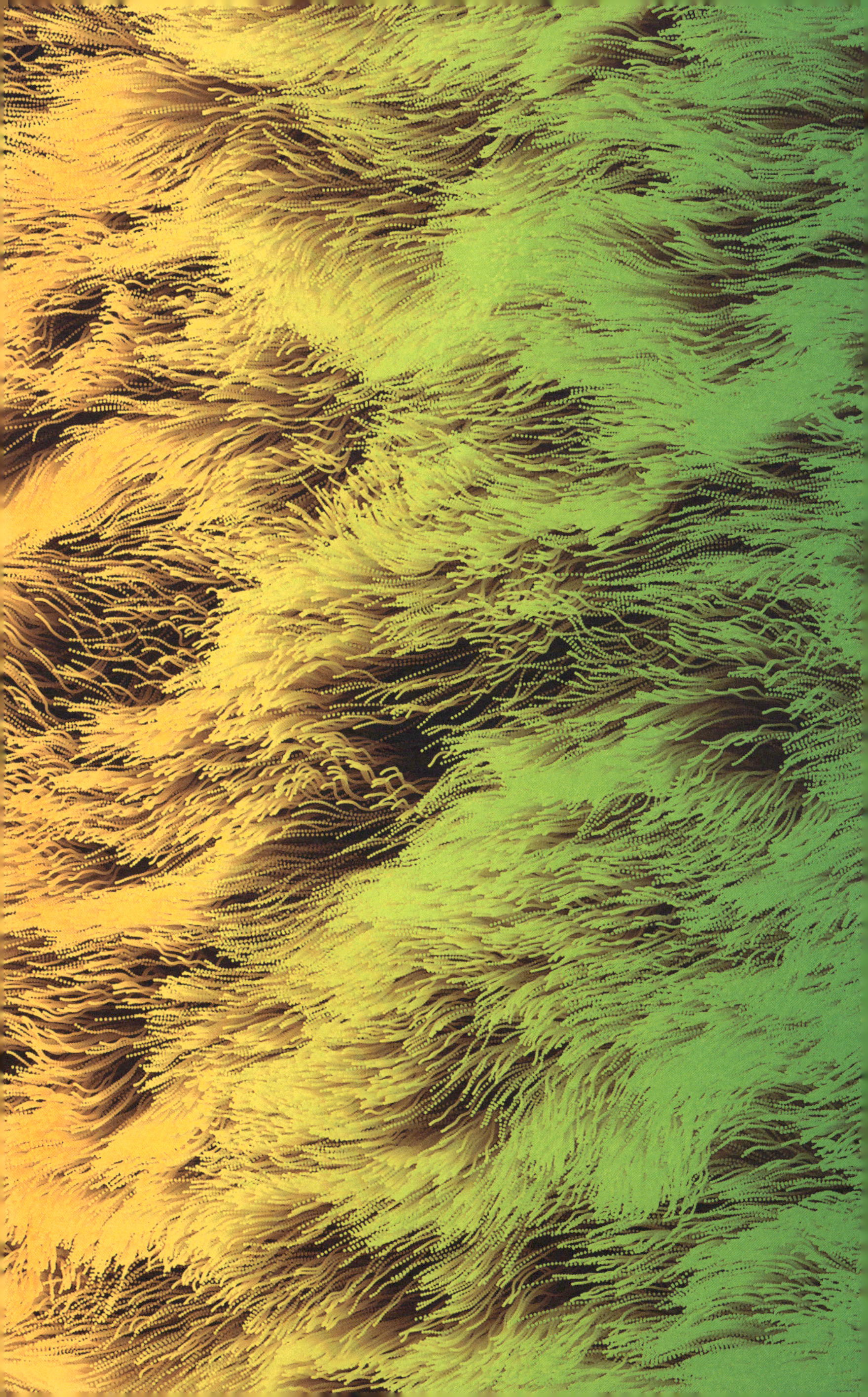

[1] Close-ups of the virtual environment intended to abstractly and joyfully express the flux of things. Image Cem Sever, Mathieu Rudaz.

[2] With the palms of their hands, visitors can influence the flow of colors sweeping across the wall. © MICR, photo Alain Germond.

← [2]

The International Red Cross and Red Crescent Museum is one of Geneva's most-visited places. It lies just next to the Red Cross and Red Crescent headquarters, a few hundred meters from the United Nations building, and offers a powerful and unique account of human cooperation. In 2011, work was undertaken to renovate and expand the museum. Roger Mayou, its director, devised an ambitious scenario entitled The Humanitarian Adventure to restructure the permanent exhibition around three topics, to be presented, if possible, by three well-known exhibition architects – each from a different continent. An ideas competition was launched and the objectives were achieved. Shigeru Ban from Japan was responsible for addressing the theme "Reducing Natural Risks." For Francis Kéré from Burkina Faso it was "Restoring Family Links." Gringo Cardia from Brazil was to tackle the subject of "Defending Human Dignity." The Atelier Oï, a fourth partner based in Switzerland, had the difficult mission of uniting these energies, creating shared spaces and bringing the project together as a whole.

For Gringo Cardia, the theme of human dignity had to find the expressive force needed to translate the reality on the ground. It had to be able to evoke the emotional charge of real situations. In one of the rooms he designed, he chose to illustrate human dignity trampled underfoot with images from archives being crushed by a giant foot. In another room, small objects made by prisoners bear witness to their precarious situation. The penultimate room – called "The Chamber of Witnesses" – encourages visitors to enter a small cabin and place their hand on a screen that displays a witness, standing almost motionless. Visitors place the palm of their hand against the witness's hand to activate the recording. The witness then begins to tell his story. Looking straight into their eyes, a child soldier tells the visitor of his journey across war-torn African countries, and how he had to kill to survive and escape. He then describes how he is trying to make a fresh start and to live with his past. For Gringo Cardia, the force of sequences like these should not, however, leave visitors with a sense of despair or powerlessness; it had to end on hopeful, dynamic note. So he formulated the idea of a final experience to get visitors involved and encourage them to act upon the course of events, but this time in a positive and cheerful atmosphere. Like the previous room, the idea was to get visitors to use the palms of their hands again, but this time on a giant visual display taking up an entire wall.

The interaction design team from the EPFL+ECAL Lab (led by Cem Sever in collaboration with Mathieu Rudaz) worked with Gavrilo Bozovic, one of the lab's two engineers, on a variety of possible scenarios (sensor systems, visualization principles, and interaction scenarios). Discussions with Gringo Cardia led to the creation of a visual environment on a tactile wall of around 18 square meters to intuitively convey to visitors a few strong, simple messages: "What we are confronted with is not inevitable. We can act and influence the course of events. And influence them even more if we work together." The initial idea was not to limit this experience to an actual specific case from the museum's archives. At the same time, it also had to avoid a symbolic approach, whose interpretations could also be particularly risky in such an international and multicultural context. On the contrary – the aim was to offer an experience that appeals to the senses, emotions and feelings to create an impression that, like the museum, fully justifies the visitor's physical presence.

After several months of technical trials and visual research, "The Colours of Dignity" project eventually got off the ground. Visitors see great swirls of abstract color spreading over the wall, inviting them to touch them with their whole hands. Immediately, the swirls of color change. This effect increases with the number of hands. The wall and its tactile system were entirely designed at the EPFL+ECAL Lab, and not without a certain amount of difficulty, as this one-of-kind installation could not justify the participation of an industrial partner to guarantee its use by hundreds of thousands of visitors in a less stable environment than a laboratory. But in the end, by pooling all their talents, they succeeded – the wall floods visitors with light, making them part of a colorful, joyful celebration to promote human dignity.

CURATOR
Nicolas Henchoz

DESIGNERS
Cem Sever, Mathieu Rudaz

ENGINEERS
Gavrilo Bozovic, Emmanuel Senft

Part Two
Visions and Proposals

The second part of this book draws on the Lausanne experience, its references, sources of inspiration and many encounters to address some of the issues that lie at the heart of innovation. Is disruptive innovation utopic? How can design unite with emerging technologies? How has its emancipation from modernism in the 1970s influenced the digital world today? Can we transcend the power of storytelling? Our goal in this second section is not to provide proofs, but rather reflects a twofold ambition: the first is to formulate visions and proposals that will aid in developing the relationship between society and those technologies that so greatly influence our daily lives; the second is to redefine design's contribution to such a challenge.

A New Digital Revolution

"Betty, a dapper 90-year old Londoner, would like to develop new contacts as most of her friends have passed away, but she isn't sure that interfaces like Facebook are appropriate because they are too generic. But Betty is curious, and explores the different options using her grandson's account," explains Peter Ziegler in "You, Me & Everyone We Know." This designer performed his research at the Helen Hamlyn Centre of the Royal College of Art, as part of a project initiated by the EPFL+ECAL Lab on social networks and the elderly. The project is part of a larger program that aims to explore our relationship to the digital world. Its main hypothesis relates to the growing complexity of human-machine interfaces, whose development seems largely dictated by the idea of performance: more features, more buttons, more commands – and faster. Ziegler's idea was to explore new approaches, making user's experience the focus.

The program began with two real-life cases. The first, "Lazy Bytes," addresses this interaction via a physical object and daily life: the TV remote control. The second, "You, Me & Everyone We Know," based on social networks, centers around the computer screen. Collaboration with the elderly was important, as it provided new and inclusive perspectives for understanding our relationship to digital technology and bringing generations together. In order to establish a body of specific data, Peter Ziegler conducted a field survey, with 33 interviews and eight in-depth sessions with people aged 60 to 90, as well as a co-creation workshop. He likewise used some other competencies of the Royal College of Art, including organizing a workshop with Jeff Willis, head of the visual communications program, and his students to devise new forms of visualization. The fieldwork and early design explorations laid the foundations for formalizing new concepts related to social networks. Peter Ziegler in London, and a second team at the EPFL+ECAL Lab in Lausanne, led by Cem Sever and Thibault Brevet, then developed concrete proposals. Throughout the project, Rama Gheerawo, head of the Age and Ability Research Lab at the Helen Hamlyn Centre, and Jeremy Myerson, the Centre's director, instilled their vision and practices based on principles of ethnology and anthropology. In the end, the project also offered a better understanding of the issues underlying our relationship to digital technology, challenging many common misconceptions and contributing to the development of the new vision proposed in this chapter.

For older people, the digital rift between generations often gives rise to the same statement: "Young people, they're born with it. It's much easier for them." But what are the cultural, technical, intellectual and physical barriers that actually create this divide? In the context of social networks, studies refute many assumptions. First of all, on a social level, they show that people do not isolate themselves with age; older people *want* to have active social lives and stay connected to society. Secondly, these studies show that the elderly can also manage complex interfaces by using various mechanisms such as selective optimization and compensation, to overcome certain physical limitations. Like Betty, our Londoner, more and more people use the current interfaces for lack of another choice. So, where exactly do the difficulties lie? Mostly in the balance between life stage and cultural context. When we are teens or young adults, the main challenge is defining our identity and existence. Social networks become formidable allies for measuring the progression of our influence and impact, as evidenced by the prized number of "friends" we have. True, we can publish anything, see anything and join different groups, but do we really know these "friends"? Is this content authentic? Will these interactions somehow translate into a reality in our lives? These questions aren't very important when creating one's identity. Social networks encourage us to develop that identity, to create mirrors, and to compare ourselves with others who are trying to do the same thing.

Conversely, as time passes, the number of professional and family commitments grows. Our address books fill up, but free time becomes an increasingly rare commodity. The race for the number of virtual friends becomes less important; the question then becomes how to cultivate real friendships despite the pressure of daily obligations – in other words, how to develop new relationships that will have an impact in real life, be it for leisure or business purposes.

For older people, Peter Ziegler highlights three obstacles: the first is capacities. Lack of education, physical disability and/or a marginal socio-cultural context can create a divide that is difficult to overcome. But this first barrier is often wrongly blamed, making for an easy interpretation. The second barrier concerns trust.

Age engenders different types of fragility and, hence, anxiety and even fear. Sharing one's interests and personal history requires a climate of trust. Finally, the third barrier is belief in the value, importance and utility of engaging in such forms of interaction. Even though people have more time after retirement, they also seem to have the impression that time passes more quickly. At the same time, the perception of futility becomes greater and often prohibitive.

It is clear from this research that our relationship to digital technology depends above all on the need for interfaces, services and practices that correspond to our life stage. On the one hand, the older we get, the greater our need to connect our virtual and physical existences becomes. Extremely independent during adolescence, these two worlds gradually combine and enrich one another. The *quantity* of relationships we have is replaced by their *quality* and actual impact on our lives. On the other hand, it is also worth noting how users can utilize services based on new technology. In 1992, Roger Silverstone and Eric Hirsch described the principles for appropriating a technological interface, incorporating it into our life context and using it to interact with the outside world. The interface design must go beyond mere physical ergonomics (with a few stylistic details thrown in), to provide a real language that resonates with the user's cultural and social context. In 2008, in his thesis entitled *Stress in Senior Computer Interaction*, Henk Herman Nap stresses the fact that good physical ergonomics improve user performance (in test conditions, users performed tasks more quickly). However, if the interface design expresses no relationship with the user, it can lead to a great deal of stress for the latter – despite its simplicity – that will ultimately discourage the user. Older people thus tend to reject remote controls with large buttons, which cruelly mirror their increasingly vulnerable condition. A similar phenomenon threatens interfaces with an overtly scientific visual identity. Despite exemplary physical ergonomics, they induce stress and are an aversion to many people who are not particularly scientifically oriented.

But it is clear that the culture of digital interfaces is still deeply rooted in a Modernist identity – that is, to maximize performance, increase the number of functions and optimize production to improve competitiveness. In visual terms, we streamline these famous physical ergonomics – undoubtedly essential – to their maximum potential. And a stylistic touch (often efficient and minimalist) creates the aesthetic of the final product. Destorting the words of Louis Sullivan, we can say Visual expression follows function! Interface design is still mainly influenced by the Modernist approach artfully epitomized by designer Dieter Rams in the 1960s (and Otl Aicher in the fields of graphic design and typography), and is often enhanced by online communities that scrutinize all of the technical aspects of the devices and compete with rational arguments to play the role of advisor.

Could not popular culture somehow use these interfaces more effectively? Could we not spark another revolution in the digital, like the one Archizoom and other Italian groups started at the end of the 1960s, defying Rationalist expression? Some evidence of such use does exist, but is often confined to an anecdotal or accessory position (like protective cell phones cases that use extravagant materials, forms and graphics, or the analogy to audio cassettes or other playful expressions (pink bunny ears, etc.). However, references to popular culture are less often involved in the design of the object. As mentioned earlier, the project "Lazy Bytes" addressed the issue by looking at the TV remote control, questioning the race for functions and technical performance, and focusing on its balance with our various life contexts and with emotions instead.

The notion of object plays a key role in a time when all of our interactions can be controlled by software installed on cell phones, or by using simple gestures (with motion detectors). First and foremost, however, the project explores the value – read necessity – of reconnecting the virtual world with the physical one, our senses, and the pleasure they provide us. We can draw a similar parallel with food. During early space missions, modernism boasted new forms of nutrition (pellets, freeze-drying, etc.) – food whose ingredients were completely controlled and so much more convenient, efficient, and manageable than a cooked meal. This utopic vision lives on. Today, cooking once again places value on the product, its history, the emotions that certain dishes evoke and the pleasure we take in sharing it with friends. The Lazy Bytes' approach is not to compare rationality and culture, but rather to unite them in the innovation process. Technological performance and its rational approach is no longer an end but a new starting point for innovative practices in their popular expression.

The Paradox of Digital Society

The idea here is to use design to define a new relationship with the digital, but also leads to a more radical questioning at the societal level. The inherently rational approach used in the designing of digital tools gradually leads to the emergence of new dogmas. And it is indeed an emergence, since it is not the result of a willful overall strategy, but rather is a spontaneous, progressive crystallization. For instance, Google is part of a study being carried out at Stanford that has made it possible to reduce each web page to a single digit – the "Page Ranks". This ultra-efficient vision has fundamentally changed our relationship to web content by developing a system that greatly facilitates the search for content. Its incredible simplicity allows anyone to utilize new content. More interesting still, it has put an emphasis on the free content and its many related services. Developing this offer, however, comes at a cost (namely research on algorithms, computer infrastructures and dividends for investors) and has therefore led to novel advertising services, an alternative valuation of content and new business models. Thus we are gradually moving towards a model that pays less for producing content than for making it accessible. In Switzerland, two thirds of the turnover of the digital advertising market goes to search engines and other web service providers, according to Roger Wassmer from the company Zanox. The concepts of privacy and identity have also become permeable, in exchange for this free use.

A paradoxical situation is gradually taking shape. On the one hand, everything seems to be becoming more free and accessible; on the other hand, new principles gradually emerge and affect how society functions. Social networks epitomize this. While they offer tools that are user-centered, where individuals can organize their environment, relationships, and interests, thus pushing the postmodern way of life (or, more generally, this society of after-modernity), these free tools have given birth to a business model that capitalizes on links and content provided by users. In the quest for efficiency and results, this model streamlines and optimizes the services. Interfaces determine our interactions, making them positive by negating part of their animality. Hence, we can "like" but not "dislike." We can become "friends" but cannot hate one another. But blurring or eradicating our animality is one of the great principles of modernism, which aimed to organize society based on great founding texts, with a lingering belief that man is the owner and master of nature. These tools thus mark the emergence of new dogmas that gradually govern interactions between individuals and cultural production and influence social organization. Social networks totally embody this paradox of apparent freedom and implicit control.

So, where will this paradox lead? Sociologist Michel Maffesoli views it with an air of confidence. For him, society procures tools, appropriates them, uses them, tweaks them and diverts them. The ups and downs can be more or less great, but all lead to a convergence. We could likewise see it in a negative or pessimistic light, fearing instead the tension of this paradox to the risk of rupture. Design can help avoid such a rupture and can even stimulate this unification by proposing new uses and developing new practices and modes of expression that better link these digital tools with societies and their cultures, to develop interfaces that strengthen the transparency of issues.

Interview with Andrea Branzi
A World Made of Exceptions

His EROS exhibition in Milan in April 2013 says a lot about this designer, architect, thinker and visionary. At seventy-five, Andrea Branzi is still an uncontested reference when it comes to tickling our neurons. EROS features his work on the erotic (erotic objects with forms suggestive of sexual attributes, expressive scenes inspired by the Kama Sutra, and so on). There's no doubt about it, this exhibition accurately captures the personality of man – not as a sexual being but one who speaks of emotion, love and the communion of bodies and souls. These are not mundane erotic scenes cut from brass or ceramic objects, but rather an invitation for us to reconnect with our bodies. EROS speaks to us of the social erotic; with it, Branzi once again expresses the "irreverence" for modern society that has characterized his career. This man has made his mark on history with his revolutionary impact. Here he talks about creating Archizoom and its manifestos to further sustain this book's focus on innovation and society today, stimulating the discussion about design's role in the digital world.

From the very beginning of Archizoom, you used icons from popular culture to challenge the certitudes of modernism. Where did this desire come from?

The Radical movement arose from an urgent need for significant realism, both cultural and political, relative to the reformist utopias of the Rationalist movement. In this respect, pop culture was an important discovery of the real world, the market and mass communications, beyond the cultural and political limitations of ideologies and the Cold War.

Did proponents of Rationalism like Dieter Rams take your proposals as an opening or an attack?

The Ulm School proposed a modernity based on scientific certainties and believed in an industrial production that was totally alien to the laws of the market. The Radical movement contested this type of Utopia. I don't think Dieter Rahms ever understood our approach as it didn't correspond to his absolute certainties.

20 years later, you were a leading member of the Memphis movement founded by Ettore Sottsass. What makes you keep going back to the drawing board? Is it a desire for distance as a kind of safeguard?

At the root of Memphis and Alchymia was an intuition that proved prophetic: that from the 1970s onwards, the mass market would gradually be replaced by niche markets. This change, which the industry wasn't aware of yet, needed new products capable of choosing their own user through the expressive power of color, décor and form. So once again, the Italian New Design, the heir to the Radical movement, was the first to understand society's transformation in a realistic manner.

Rationalism (and hence modernism) always come charging back, because they are inherently reassuring for the entire decision-making chain in a company. Should the pop culture revolution be permanent? What is the designer's mission?

Classic modernity believed that industrial logic, factory organization and production symbolized the future of society as a whole, and that the Industrial Revolution guaranteed a future of order and reason. So the factory was society's model. The Radical movement believed the opposite – that society should be the model for the factory. What you call "popular culture" is a complex, contradictory and constantly changing reality. Design does not have an independent mission but a critical role in the history of society.

> The design of digital interfaces for computers and the internet is decidedly Modernist. Is there any way of envisaging a revolution in this area?

An IT revolution is not just about technology or design, but a phase of anthropological evolution – a new human evolution wherein memory is contained in hard drives, orientation is contained in browsers and information is contained on the Internet…a fascinating new frontier in the history of mankind.

> We talked to Jasper Morrison about his idea of "Super Normal." From a visual point of view, it's very minimalist – the antithesis of Archizoom's proposals (such as gazebos). But, fundamentally, it's also a radical proposal in relation to popular culture. What do you think of the idea of Super Normal?

Normality doesn't exist today. Perhaps it never has. The world is made of exceptions, variations, unforeseen events, and constant innovation. There is no minimum, moral or political certainty. Reformism no longer has any direction to head in.

> From the end of the 1960s, counter-culture movements strongly questioned what was seemingly established. What do you think of today's design?

I think that contemporary design is too hedonistic, self-referential and is a throwback to the last century. It's completely removed from the reality of the 21st century. It should start tackling the great cultural themes like life, death, the sacred, Eros, etc. – realities that classic modernism has totally ignored.

During your career, you've made a permanent mark in the history of design. Which of your actions or approaches do you think were the most relevant and still have the most impact today?

I think Non Stop City represents the project's main hypothesis in the age of globalization, especially through its enigmatic and radical vision of the city of the future, without limits.

Project

Lazy Bytes

+ VOL –
+ CH –

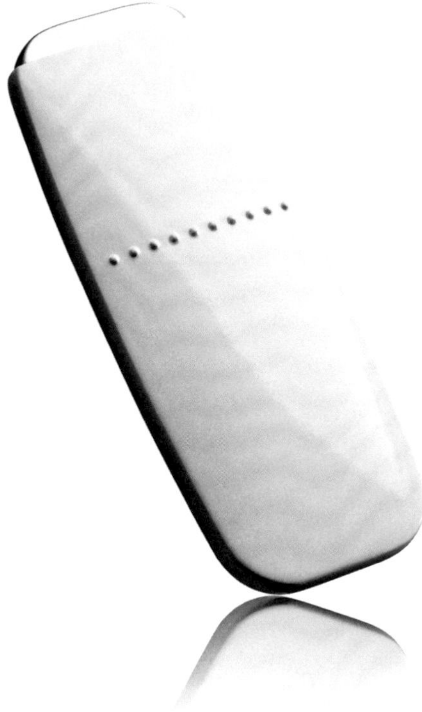

[2]
[3]

Why do we invest our money in a vase, a lamp, or a chair, but not a TV remote control, also a central feature of our living room? Several great designers like Philippe Starck, Naoto Fukasawa and Yves Béhar have approached the question with interest. But the revolution has not yet occurred. Sixty years after its creation under the nickname "Lazy Bones," the remote control seems to remain nothing more than a simple set of buttons and functions embedded in a block of plastic, eliciting no emotion, no aesthetic principles, no memory. In short, it has not yet truly acquired the status of an "object" in and of itself. And yet, it is a vital daily link between our bodies – which are real – and access to the digital world, of which television is now a part.

The EPFL+ECAL Lab's aim is not only to open up new prospects for specific technology, but also to provide answers to practical questions raised by the use of these technologies. Our relationship to digital content is part of this. How to renew this relationship that, today, is still largely driven by a race for performance, increasing the number of functions and reducing prices? This race, however, excludes a great many users, including the elderly and, more generally, those who have no particular attraction to technology and/or feel no need to master it. Two concrete cases were used to illustrate this problem: social networks in relation to the elderly (representation on the screen) and the TV remote control (exploring the object as a physical interface). The latter gave way to the project "Lazy Bytes," whose aim was to break with the quasi-acneic development of the descendants of the original Lazy Bones.

The Lazy Bytes project arrived in a novel context in that it did not intend to replace the most recent generation of highly-sophisticated remote controls, which were designed to handle the entire home environment and advanced programming features (which was now the domain of touch screen tablets). It could, therefore, instead focus on the idea of offering something extra, something different – a new sensory and cultural experience that would renew our relationship to digital technology in daily use. Four top design schools – l'ENSCI-Les Ateliers in Paris, the Royal College of Art in London, Parsons The New School for Design in New York and the ECAL/University of Art and Design, Lausanne – agreed to participate in this challenge. For an entire semester, more than 100 students, teachers, designers and engineers helped develop concrete proposals. The goal came from the strength and diversity

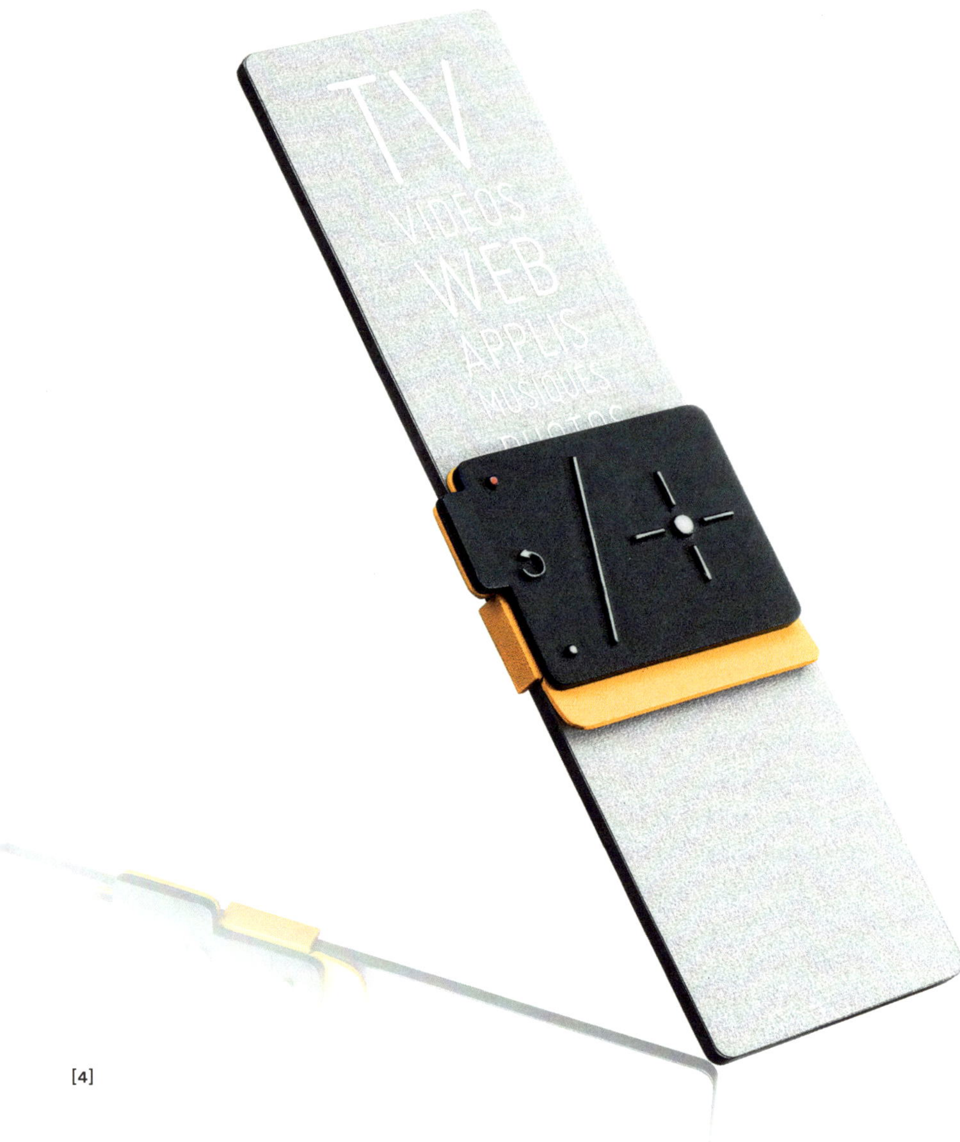

[4]

of a single approach, as all participants started from the same point. The framework laid out by the EPFL+ECAL Lab allowed the actors to choose the level of complexity of their project by defining a coherent set of functions that their remote control would perform.

The Kudelski Group, a global leader in access control to TV channels, offered its expertise by presenting the current challenges of digital television and, more generally, the battle among equipment suppliers, content providers and service providers to conquer users' living rooms. The EPFL+ECAL Lab laid out the specifications with regard to the design, building on recent research on the notions of perception and ergonomics. Ergonomics indeed necessitate a scientific approach involving physiological and biomechanical elements, as discussed in the previous chapter. However, Lazy Bytes had yet another goal: to create an emotional, stimulating, and sensitive relationship between the object and its user, thus helping to reduce the latter's stress and strengthening his or her desire for appropriation. Support and interaction that targets several generations in a single interface is also an important criterion for success and, consequently, a field of action for the design field. Thus, the Lazy Bytes challenge also consisted in designing remote controls that did not stigmatize a certain group of consumers.

The designers had the freedom to highlight the qualities and values of the object itself. Hence, their exploration focused on potential uses as well as visual language, materials, gestural quality, tactile sensations and, cultural references. Within such freedom and openness, each school developed a specific strategy. A simplistic overview of these approaches might cite ECAL's project's (under the direction of Rémy Jacquet, and Laetitia Florin) communicative strength, while that of Parsons (under the direction of Zach Libermann, Amit Pitaru, and David Carroll) questioned potential interaction scenarios. The Royal College of Arts (under the guidance of Martin Postler and Ian Ferguson) sought new opportunities, while the ENSCI-Les Ateliers, which (with Laurent Massaloux) chose to develop all of its courses for the entire semester around this challenge, called upon several speakers from complementary disciplines (i.e. notions of anthropology and sociology combined with an in-depth knowledge of technological possibilities).

Some students played on the complementarity of existing complex remotes by limiting their proposals to the most basic set of functions (changing channels, adjusting the volume and an on/off function), like the small "Freehand" remote, made of colored silicon, which can be attached to any object via two small suction cups (thus turning *it* into a remote control!). Others chose to include a much wider range of functions by concentrating on potential interaction scenarios. With the "Circule" remote, users can actually control all of their living room's devices via a central joystick and mini-screens. Lazy Bytes resulted in 63 projects in all (28 of which are in the show and are featured in the catalog), with tangible benefits, including a number of patents and trademarks.

Scale models, some of which were made into functional prototypes, were used for a series of tests. In the end, and in light of the new proposals, none of the participants wanted to keep their old remote. Three-quarters wanted to use a remote they had selected from among the Lazy Bytes remotes, while the other quarter would have liked more information about how it worked before deciding. The "Touch Speak Walk" remote, which combines vocal and tactile interaction with great simplicity, garnered the most votes. More importantly, projects like Galeo (a magnetic roller on a wooden base) highlight the object in complement with digital technology. Lazy Bytes' findings far surpass the development of specific models or principles; they provide an important basis for the development of new interfaces that connect the everyday physical world and digital content.

CURATOR
Nicolas Henchoz, EPFL+ECAL Lab

SCHOOLS AND INSTRUCTORS
Royal College of Art:
Martin Postler, Ian Ferguson, Tord Boontje.
Parsons The New School for Design:
Zach Lieberman, Amit Pitaru, David Carroll
ENSCI-Les Ateliers:
Laurent Massaloux, Romain Cuvellier, Jacques-François Marchandise
ECAL/University of Art and Design, Lausanne:
Rémy Jacquet, Laetitia Florin, Elric Petit

ENGINEERS
Olivier Nguyen, Gavrilo Bozovic, Baptiste Junker

INDUSTRIAL DESIGNERS
Anne-Cécile Rappa, Susanne Schneider, Gregory Syrvet

TEST USER
Pascal Magnenat

SCENOGRAPHIE
Adrien Rovero Studio

IMAGES
Olivier Pasqual

INDUSTRIAL EXPERT
Thierry Dagaeff, Kudelski Group

INDUSTRIAL PARTNER
Kudelski Group

SUPPORT
Fondation Leenaards, Loterie Romande

[5]

[6]
[7]

[8]

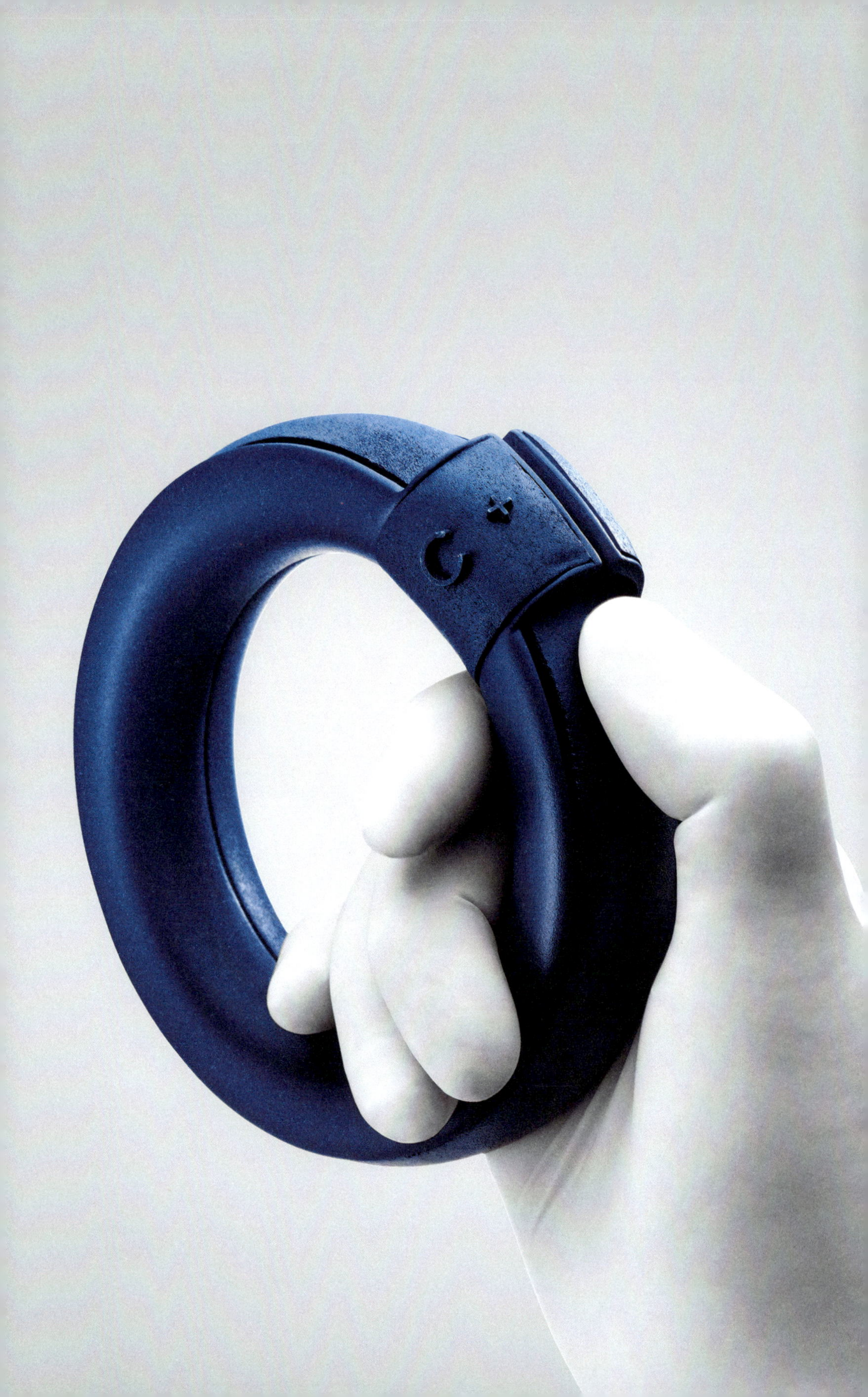

[1] GALEO
Raphael Dutoit
ECAL

The silky texture of this hardwood base and black composite material roller command completely renews the material codes and feel of the remote control. The roller attaches to the base magnetically.

[2] TOUCH SPEAK WALK
David Bellisario
RCA

No buttons here. Instead, the user browses using a touch pad on the white top part. The volume is adjusted by sliding one's finger along the dotted line. Vocal recognition completes the functions. The user confirms the instructions by squeezing the object's soft covering with his or her hand.

[3] ZAP
Marie Schenker
ECAL

This book has been printed with conductive ink, ensuring contact between the pages. When a user opens it, the television shows the number corresponding to the double page that is open. The paper's surface allows the user to add numerous handwritten notes. The on/off and volume functions are located on the book's spine.

[4] PLATFORM
Florent Julien
ENSCI

A small sliding command platform on a base identifies the digital device to control. The introduction of this movement greatly simplifies the interface, which is reinforced by the overall graphic design work and the platform's reliefs.

[5] CIRCULE
Arthur Siau
ENSCI

Objective: to control with a single joystick all the digital interfaces within a living room. Five miniature screens located around the joystick display the necessary information to make navigation intuitive.

[6] ROLLING CONTROL
Mathieu Rohrer
ECAL

Inspired by the games of yesteryear, this device allows the user to control 999 channels rapidly, simply by turning three faceted dials corresponding to hundreds, tens and units. The smallest disc (front) allows you to control the volume.

[7] À PORTÉE D'ONDES
Emile Kirsch
ENSCI

This base serves not only as a remote but also a speaker, so that you can take the TV's sound with you wherever you go – even far away! The central part houses lozenges that allow the user to launch different content and stimulate discussion among family members. The project is based on anthropological research.

[8] TWISTYMOTE
Hilal Koyuncu
Leif Percifield
Francisco Zamorano
PARSONS

A turning handle with four colors matching four possible customizations. The project offers a complete usage scenario, with automatic learning of user preferences.

[9] IRMA
Zoé Aegerter
ENSCI

An overhaul of gestural interaction: the orientation of the ring permits the selection of a channel from a mosaic displayed on the television. The outer surface of the ring is a rotating band used to zoom in on the mosaic to select content or functions.

[9]

Original and Super Normal

In the 1980s, a young radio station by the name of Couleur 3 dared to poke fun at itself in its ad: “Couleur 3, the fourth noisiest thing after the coffee maker, the dishwasher, and your neighbor’s kid.” Today, this formula provides a source of inspiration for the challenge of innovation by expressing the need to relativize the importance of novelty in the flow of information and disturbances of daily life. Users hardly have time to understand the promise of a new product with a revolutionary ambition; if they are not being disturbed by the lawnmower or the “neighbor’s kid,” they are overwhelmed by the stress of work, next year’s tax return and the colossal sums of money spent defining their social status! And yet, innovation fosters real ambitions in these already overworked users. For sociologists like Norbert Alter, “innovation differs from invention in the sense that it represents the implementation of this innovation and its integration in the social setting... If the social body appropriates it, the invention becomes innovation, and the effects of its implementation are numerous.” But, as users who are already completely overwhelmed, we cannot be expected to make a major effort to understand and learn new visual codes or question the values that control our daily lives; innovation must naturally fit into our social and cultural context.

This becomes even more complex with disruptive innovation, which does not lead to a known solution but instead offers a radically new proposition: it is disruptive, and therefore “in rupture” with that which came before it. So, the question is, how to make it a natural part of users’ daily lives? Is not the term “disruptive innovation” more than just a contradiction in terms? Is it not an oxymoron? Literally a utopic idea doomed to failure by definition? The first step is to decouple technology, use, and contextualization, to reform it so that the question then becomes can we build on radically new technology to create novel practices that will naturally become part of its users’ social and cultural context? The role of design and its potential contribution when it comes to exploring these uses and their contextualization then becomes clearer. Among the main sources of inspiration for meeting this challenge is the exhibition “Super Normal: the Sensation of the Ordinary,” created by Japser Morrison and Naoto Fukasawa in 2006. The two designers elaborated on their approach in a book the following year.

Naoto Fukasawa sums up the idea behind their proposal. "Normal is a situation where something has blended comfortably into our lives. I believe that Super Normal is about extracting the essence of [this] Normality." Despite the minimalist feel that might emerge from this exhibition and the number of objects referenced, Jasper Morrison denies any purely functional approach or any modernism in the exhibition catalogue. "In the 70s, design was all about function, but it was the function of the moment – the kind of function you test while opening a can of peas, asking yourself in that moment how good the can opener is. These days we take function for granted, and for the most part things work well enough not to complain about them. So taking a pure functional approach to design would be disastrous. Super Normal's about how things work in relation to our living with them." The Super Normality presented by the two designers is therefore limited neither to technical function, nor to physical ergonomics. It likewise addresses users' perceptions and their relationship with their socio-cultural context. Naoto Fukasawa and Jasper Morrison deny wanting to elevate their vision to the status of design theory. Instead they seek to offer a new interpretation of our relationship to objects and the appropriation of new ideas.

Such a view is an important means of shattering this apparent paradox of disruptive innovation and clarifying the roles of the disciplines involved. Scientific research and the technologies emerging from academic laboratories should have the freedom to make disruptive proposals. As radical and new as they may be, these proposals are nonetheless based on the strength of a rational approach proven through reproducible experimental verification. Again, in an open academic setting, design can explore new use scenarios, which can also result in a break with the existing and lead to new practices. It is likewise responsible for developing a new, often visual language that creates bridges with the everyday lives of users. The Post-Modern revolution of the 1970s gave it the freedom to approach this role in an appreciable manner, to fearlessly collide with the logical proposals emerging from the field of engineering through a vision based on diverse popular cultures. Design must no longer limit itself to the role of "director" of new features or performance levels, as Roger Tallon called for;

rather, design can reinterpret them, giving them new meaning and instilling them with veritable capacity for expression.

In the late 60s, while Archizoom and several other groups of designers and architects with revolutionary ambitions led the charge against Rationalism with "eclecticism and afro-tyrolean kitsch," others (like Ettore Sottsass and his Valentine, or Alessandro Mendini and his language based on form, which also incorporates popular references in mass-produced objects) espoused a constructive stance, to actually become part of the innovation process. These designers changed the way we live with the objects they created. The decades that followed consecrated this contextual, sensitive, decidedly Post-Modern approach, but mostly for simple objects and rarely in relation to emerging technologies. Digital technology still lacks such a contribution, as we saw in the previous chapter. While such movement took place in the artistic and theoretical spheres, where players jousted with revolutionary terms to give meaning to new movements, it hardly affected products from the marketplace. Another example is the fact that the potential of new materials in everyday objects remains largely unexplored; today they are often used either for technical purposes, or to give objects a "high-tech" look. The project "Hidden Carbon" (presented in this book) addresses this very topic. In this project, carbon was used to offer new experiences with respect to appearance, texture, and interaction with light, rather than as a mere selling point.

Nothing to Hide

This quest for normality does not necessarily indicate a systematic attempt to hide any technological element. As the augmented reality research in the project "Gimme More" shows, reducing the technological barrier is strongly linked to the user's perception. The installation created for Camille Scherrer's *Le Monde des montagnes (World of mountains)* does not attempt to hide the screen or the computer; the elements are presented in a simple, sober way, like an ordinary office table, the likes of which we might find in anyone's home. So, in spite of the machine's visible presence, the installation proves more convincing than Maria Laura Mendez Martens' "Dots" project, whose installation uses a more discrete but less common screen. It attracts the eye, thus intriguing the viewer and putting the technology at the forefront.

The idea of Super Normality – the source of inspiration for facing the challenge of disruptive innovation – has obvious links to the works of Hans Henrick Nap mentioned earlier. The researcher analyses the key elements to establish a relationship of comfort and normalcy between a user and an object, as Naoto Fukasawa stated at the beginning of this chapter – elements that go beyond the notion of physical ergonomics to address questions of identity and culture.

But can we learn from Super Normality, without trivializing disruptive innovation in the existing? This comes back the relationship between Super Normality and the successful revolution carried out by Italian designers starting in the late 60s. Both, however – each in their own way – made design part of the innovation process and helped define its role in turning disruptive technologies into uses with regard to peoples' cultural context. Morrison and Fukasawa mainly drew their findings by choosing existing, seemingly mundane objects (as they are already integrated in our everyday lives) to illustrate their point. This partly explains the feeling of minimalism. Sottsass, Mendini, and their colleagues, on the other hand, took part in a revolution and formulated proposals precisely *to change* the existing relationships between the object and user. Naturally, their proposals were more spectacular and provocative. Both examples, however, call for an approach that can transcend the apparent paradox of disruptive innovation ('apparent' because, in the approach taken by the EPFL+ECAL Lab, this paradox does not exist). There is a strong need to reorient the research and exploration at the interface between the fields of engineering and design, and, more fundamentally, between two worlds that represent very different cultures and approaches. As Michel Maffesoli stated, the challenge lies not in the mere struggle, but in the need to go beyond traditional approaches to generate the emergence of new ones.

Interview with Jasper Morrison
Living with the Object

Very British, but now mainly based in Japan, Jasper Morrison is a man of few words. However, in the middle of the Hermès presentation at the Milan Furniture Fair in mid-April 2013, when we talked to him about the EPFL+ECAL Lab and the influence of his book, his eyes began to sparkle. *Super Normal: Sensations of the Ordinary*, co-written with Naoto Fukasawa, indeed offers important reflections with regard to rising to the challenge of disruptive innovation, which is both radically new and, at the same time, connected with society. Normal describes a situation where something has comfortably blended into our lives, hence "becoming normal." "Super Normal" is about extracting the essence of this Normality. To understand more about Morrison's philosophy, you can visit the homepage of his website, which could not convey this idea more clearly: an interior with black, stylized furniture and objects on a white background, which instinctively takes visitors to the section they are browsing over. This is pure Morrison – simple, effective, intuitive, and user-centered – not a sterile feature but a lively one that unites users. This Englishman, a graduate of Kingston Polytechnic Design School and the Royal College of Art, has been working for prestigious manufacturers such as Vitra, Cappellini, Alessi, Canon, Rado, Galerie kreo, Flos, Magis and Muji since 1986, when he set up his studio there. In 1995, he even designed the Hanover tramway. He was awarded the title 'Royal Designer for Industry' in 2001.

It's been several years now since your exhibition and book with Naoto Fukasawa. How do you see your Super Normal idea?

I'm delighted that it's getting some attention, because I still think it's the way of getting design out of the "boutique," away from what the media would like it to be: a form of distraction, a conversation piece that you should have in your home. I still appreciate some objects that seem to have Super Normal qualities, and always try to design objects that create a good atmosphere. Super Normal helps me stay closer to people by designing objects that will be useful for as long as possible.

The visual expression of Super Normal is quite the opposite of the iconic object of the Post-Modern revolution, like the Archizoom Safari Sofa. But at the same time, each of you introduced a very strong connection with popular culture. Would you say this is a contradiction or a shared vision?

It might be both a contradiction and a continuation of the expectations that designers periodically have of making design popular again. But, as always, it involves an elitist position. The difference is probably in the end product, which, in the case of Super Normal, results in "something more acceptable to live with." If the consumer can accept that, then normal is better than special.

Although you say in your book that the Super Normal is not a theory or an ideology, how does it affect the process of designing an object?

It's very simple. You just have to observe everyday situations and identify objects that outshine or dominate the others, try to understand why, and learn their secrets.

You say that familiarity is important. But don't you run the risk of recreating some of the archetypes derived from the collective imagination?

“Familiar” doesn’t mean identical. There is a process involved in bringing new life to this “familiarity,” and these objects must be better than their “ancestors”.

> Super Normal has been a permanent reference for the EPFL+ECAL Lab, especially when it comes to exploring digital interfaces and interactions. Most of the Lab’s work tries to create a disruptive experience for users, but also wants to achieve some form of normality that will make the finding legitimate and a part of the everyday. Can Super Normal be a tool in the design of media and interaction?

That’s a difficult question for me, in the sense that I don’t know much about your work. But my feeling is that Super Normal can be a real help in a variety of activities.

> More generally, can the idea of Super Normal be an asset in innovation? Should it be involved in the innovation process?

I think anything that reduces the emphasis on and importance of form in design is positive, especially in order to develop an object’s other qualities. Innovation is more interesting when linked to other aspects of design than just form. When you stop focusing on the need to propose new forms, this leaves room for other qualities in the conception process to emerge. The most successful objects are those in which form is a minor consideration, or at least does not limit other considerations.

Project

Montreux Jazz Heritage Lab

[1]

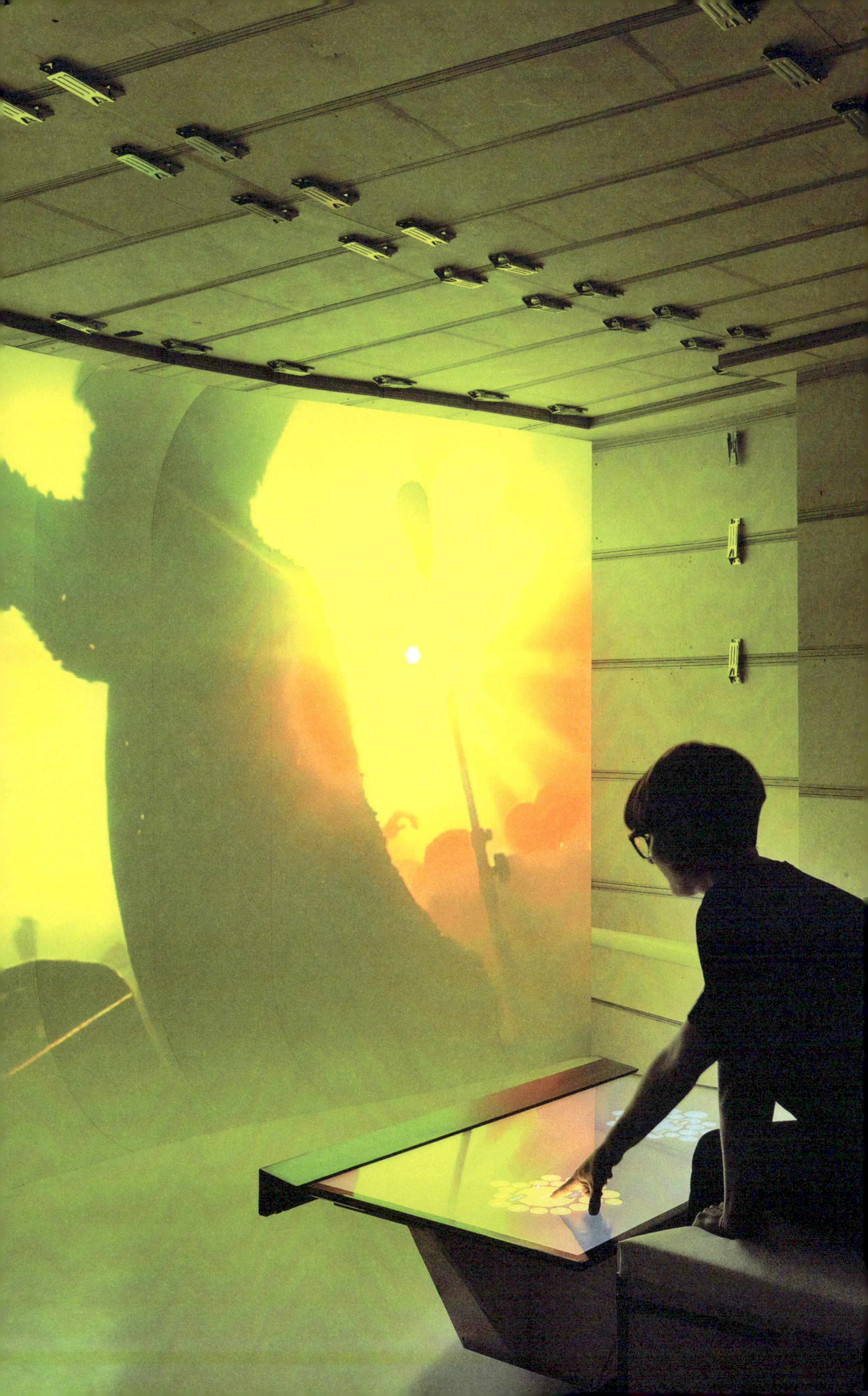

[2]
[3]

A small winding road rises above Lac Léman, gradually revealing a full panoramic view from above the heights of Geneva, all the way to the snowy peaks of the Swiss Alps. We skirt the Grand Hotel of Caux, which, in the early 1900s, was the high place of romantic tourism, where the likes of Rudyard Kipling and Lord Byron sojourned. A few more turns and we stop in front of a discrete entrance to a chalet, nearly 1,400 meters above sea level. Behind the dark wooden door resides one of music's legendary venues – the chalet of Claude Nobs, founder of the Montreux Jazz Festival which debuted in 1967. Far from the festivities that invade the city's lakeside wharves and concert halls every July, the chalet welcomes musicians, providing them a unique experience far from the constraints of celebrity life – an opportunity to share a *raclette* and bottle of wine from a local vineyard and relax and chat amongst fellow enthusiasts. This house also conceals a treasure. It is here that the 5,000-odd hours of audio-visual archives from the different festivals reside. From the beginning, Claude Nobs had as many concerts recorded as possible, using the best standards of the times. These exceptional archives now belong to UNESCO's world heritage. But how to preserve them – and make them accessible?

The celebrated rock group Deep Purple contributed to the Festival's notoriety by composing its famous song "Smoke on the Water" following the fire that burned down the Montreux Casino in 1978 during the event itself. Such a disaster at the Chalet would be the doom of the world's largest collection of live concert archives of jazz, blues, and rock. How to transfer this content to a digital medium while preserving its original integrity in perpetuity? Given the number of quality recording standards required, the question began to seem like a scientific challenge. In 1991, for instance, the Montreux Jazz Festival, along with Sony, made the first high definition recording in Europe. The Festival then partnered with the Ecole polytechnique fédérale de Lausanne to capitalize on the technological advances in the field. For the EPFL, this, in turn, was a wonderful new testing ground for the emerging technologies of the schools' laboratories (correcting time alterations, augmenting information about the content, isolating solos, etc.). Several laboratories got involved in this project, coordinated by the EPFL's Metamedia Center, which also oversees the digitization of thousands of hours of archives by an outside company.

One fundamental question remained: how to bring this amazing heritage, now stored on digital media, back to life? It was this question that was at the heart of the EPFL+ECAL Lab's mission – namely, to transform technological performance into an experience for the user. However, this requires a great many other skills as well. EPFL's Space Conception Research Studio (ALICE), directed by Professor Dieter Dietz, became involved in the project to initiate research on the physical and architectural dimensions. From the very first sessions, the Electromagnetics and Acoustics Laboratory proposed ideas to make sound the focus of the project. Other laboratories and EPFL start-ups likewise contributed knowledge and technologies to enhance this approach.

Given all of these potential skills, the project designers then had to determine what experience they wanted to offer the user. Should it stay as true to the original as possible? The digital world often tries to mimic the physical world. However, such ambitions are but pale imitations, because the real experience begins with the individual's choice of the concert they wish to attend, followed a few weeks later by their arrival at the site, in the festive, sizzling ambiance of an auditorium crowded with 3,000 people, all sharing a single experience – the physical presence of the musician and the uncertainty of what might occur. Attempting to reproduce a concert, staying as true to the original as possible using digital archives, could only create a stunted version of the emotional richness of a live performance.

The Montreux Jazz Heritage Lab project therefore had to free itself of this constraint and dare to create a unique status for this digital content – one complementary to the original concert. What are the specificities of such archives? For starters, the diversity of the existing material (different formats, qualities, recording styles, etc.) Today, it is even possible to record in 3D format, with a resolution four times higher than that of high definition. But bringing out the value these archives meant taking into account 5,000 hours of existing material rather than simply focusing on a few recent attempts. The issue, therefore, was not the race for definition but the quality of perception, which are two different ideas. The second distinctive feature was that the archives' value was not limited to their musical quality – as great as it may be – but likewise reflect the popular culture of an era, attest to social movements and evoke personal memories. Thus, the device also had to allow users to express and share their knowledge. Thirdly, users' physical relationship to the content and their own environment plays a key role. Most online digital archive systems are used in a solitary way, whereby users find themselves alone in front of a flat screen, using only their fingertips to start a song in between a chat and a couple of emails, or on home cinema systems, which involves no interaction whatsoever. This type of viewing is dematerialized and physically sterile.

Baroque Inspiration

The ALICE team, directed by Professor Dieter Dietz, focused its research on the shape of the screen, its position and its meaning, specifically with the idea of eliminating this milieu of solitude and lack of interaction. It explores not only the relationship between the image, the user, and the viewing space, but also the relationship to the larger context, in order to create a transcendent installation capable of fully conveying the energy of its content. The project advanced in close collaboration with EPFL+ECAL Lab designers, which explored new forms of interaction with this content to stimulate the interaction between participants and to develop more intuitive and more direct navigation. No question of competing with the classic interfaces of advanced research: the goal here was to bring music lovers together.

After a year of research, a first device tested the usage scenario: two users we able to search simultaneously using a kind of touch coffee table in a lounge-type atmosphere designed to elicit discussion, reminiscence and debate about a particular song, concert or musician. The device did not utilize "search" fields, as found in classic tools, but rather simple colored shapes to highlight the names of artists and their works. Once the user has made his or her selection, the concert comes to life on a giant screen very close by. The ALICE team drew its inspiration from the trompe l'œil shapes of Baroque churches, to create a curved screen, giving the viewer an unprecedented sense of closeness to, and depth of, the stage.

During the months following the trial, the teams refined solutions and then set the overall budget. A spectacular work of construction [4] → was carried out by a team of architects, most notably Thibault Smith, then but a third-year student. On June 19, 2012, the press attended the official opening of the first prototype of the Montreux Jazz Heritage Lab. The unit could accommodate six people in its

closed, acoustically-insulated configuration, or many more by opening one of its two lateral walls. Thanks to a system designed by Illusonic, a start-up from EPFL's Audiovisual Communications Laboratory, the quality of the sound diffusion was very high, even for those near a wall. Behind the screen, certain parts of the shell, made with acoustic glass, allowed a few image fragments to escape, thereby radiating its energy around it.

"It's amazing. I've never seen my concerts like this," Claude Nobs stated during his first visit. The greatest of compliments. The Montreux Jazz Heritage Lab does not provide copies of concerts, but rather has become a unique experience in its own right – one that values the Festival's great heritage in a specific way. It is an experience where users share memories with other users, bent over the coffee table, before getting caught up in a concert where the musicians seem so close that their eyes, movements, and complicity are palpable, thanks to the subtle choice of images made for nearly half-century by the directors of the Festival. The screen and keyboard have disappeared to become a complete spatial experience, where digital technology couples with physical space to radiate to the outside world.

Research continues to devise new contexts, including open public spaces. A Montreux Jazz Café located on the EPFL campus aims to host a new version of this project and new immersive experiences. Another program explores the link between the physical viewing device and the theme of the archives. Can we enhance sports events – where the geometric accuracy of the playing field certainly allows for less image distortion using similar methods? Many questions remain open. But this case study asserts that digital technology can feed other ambitions rather than simply copying the real world; it can create a special status of its own to provide unique social and cultural experiences.

INTERACTION DESIGN AND COORDINATION
EPFL+ECAL Lab: Nicolas Henchoz, Cem Sever, Daniel Tamburrino

ARCHITECTURE
ALICE, EPFL: Dieter Dietz, Olivier Ottevaere, Charlotte Erckrath, Lukas Lenherr, Tibo Smith

ACOUSTICS
LEMA, EPFL: Hervé Lissek, Cédric Monchâtre

OVERALL COORDINATION OF THE MONTREUX JAZZ DIGITAL PROJECT
MetaMedia Center, EPFL, Alexandre Delidais

OTHER LABORATORIES INVOLVED
Audiovisual Communications Laboratory
Signal Processing Laboratory 2

TECHNOLOGY PARTNERS
3D audio system: Illusonic (Lausanne)
Loudspeakers: Relec (Yverdon-les-Bains)
Acoustic glass: Glas Troesch (Bützberg/Bussigny VD)
Interactive table: to-fuse (Zurich)

SUPPORT
Vasily Shahnovskiy
Theresa Rydge
Audemars Piguet SA
Lombard Odier Darier Hentsch & Cie

[1] The interface plays on a simple representation of the content to privilege the relationship with the music and exchanges between users, encouraging the latter to adopt two positions: one leaning forward (stimulating interaction), and the other leaning back against the backrest (enhancing immersion into the content).
Photo Tonatiuh Ambrosetti, Daniela Droz.

[2] The entire unit serves as an interface, radiating the energy of the archival content. It can operate either closed, for optimal sound, or open, to accommodate a larger audience.
Photo Joël Tettamanti.

[3] The screen's form is inspired by trompe l'œil geometry from Baroque churches in order to lend a sense of depth to the videos, even for the oldest ones.
Photo Joël Tettamanti.

[4] The module functions both when closed, to optimize the sound, or open, to receive a larger public. Photo Joël Tettamanti.

Towards the First World

What could facilitate the general public's adoption of a radically new idea, service, or object – especially one linked to a disruptive technology? To readers seeking a magic solution in this book, we should immediately confess that the experiments conducted by the EPFL+ECAL Lab lead in a different direction. They show that we must take certain principles into consideration based on our perception, rather than have people believe in the existence of easily applicable tricks.

In the 1990s, to position a brand, sell a product or convey a political idea (and thus win over the final buyer or voter) we moved from rational argument to a more emotional method focusing, in particular, on *how* we attract a person's attention to arouse a desire for change. This is known as the 'storytelling revolution'. "From logos we move on to stories," writes Christian Salmon in his book *Storytelling: Bewitching the Modern Mind*. "It is the advent of consuming as the only way of relating to the world. Brands are claiming powers that we previously sought in drugs and myths." The act of consuming has thus become "an exercise in communications, or even planetary communion."

This practice no longer values the quality of the argument *itself*; rather, it ties it into the building of a "good story" that transforms a product, service or idea into a total experience for a target audience. In the past, marketing was about promotion. Now, it has to use brands to accommodate the consumer's vision of the world. This approach can be interpreted in very different ways. In a negative light, it can be translated as the art of creating fictitious tales to give an illusion of identity and emotion to a basic element. In a positive light, it can be associated with the capacity to express real qualities and values to connect them to the user's cultural and social background. In the second case, we can see in an analogous way a dimension of the fruitful relationship between engineering and design. In other words, the rational approach guarantees a new technological performance that contributes to a more comprehensive experience defined by a story.

The One Laptop per Child venture launched by Nicholas Negroponte, the founder of MIT MediaLab, with the participation of the designer Yves Béhar, is a good example of such an approach. It began with a great idea – to create a new generation of laptop computers to increase children's access to knowledge in educationally-deprived areas. It used disruptive technologies, partly to reduce the cost of the equipment, and also to ensure the transmission of information in areas where modern telecommunication infrastructure does not yet exist. The project therefore included a principle of self-organized networks, where each computer is not only a terminal but also a relay for the others. The machine becomes one of many nodes in a giant network. Finally, from its earliest stages, One Laptop per Child had a major design mission with regard to users, with a key question: how to design an inexpensive, stimulating machine that does not stigmatize the precarious situation of its users? As early as 1983, Nicolas Hayek had already met a similar challenge with Swatch. He did not create a watch for those without the means to treat themselves to a more luxurious brand. No! Instead, he gave Swatch its own identity, values, and relevance – independent of its owner's social status! Obviously, for One Laptop per Child, the problem was more complex. There was the design of the object and the interface, as well as the relationship with populations with very different traditions and cultures. There were also considerable technical constraints. But in the end, the computer became part of a history with essential democratic values, just like the Swatch, which translate the ambitions Victor Papanek expressed in his book *Design for the Real World* (which sent shockwaves through the design community in the 1970s) into a contemporary and efficient vision. One Laptop per Child's success does not lie in the several million users of the distributed computers but in the overall evolution of the entire IT sector. The industry has been forced to recognize that it is possible to produce far more accessible machines, in terms of both mobility and price. This way of incorporating a project in an overall history that expresses values and creates meaningful connections with the user is a major source of inspiration for the Lausanne laboratory. Yves Béhar and his San Francisco-based studio Fuseproject have made it one of their trademarks. His projects often make use of new technologies, for example the Leaf Lamp for Herman Miller, now part of MoMa's permanent collection, the Sayl Chair for the same brand, or audio systems such as the Jawbone earpiece or Jambox speakers.

How to use the idea of storytelling to foster innovation? One has to step back from the latest definition and resulting latest trend, and look at the narrative quality of a project in a broader way. The widespread use of storytelling is often associated with

American politicians. Bill Clinton is the first to spring to mind. "With him, storytelling stopped being a mere communication technique to become a driving force for American governance and leadership," Olivier Clodong and Georges Chétochine explain in *Le Storytelling en Action*. Back in the 1980s, Ronald Reagan was already using this technique to justify Star Wars to the American people and the rest of the world. Many earlier examples also make use of the emotional principles of narration to gain the support of the general public. For example, the identity of modern-day Switzerland is partly based on the creation of a romantic myth in the mid-19th century: relics discovered in the lakes combined with the imaginary of the times, giving rise to a valiant, prehistoric, lakeside civilization set in a wild, idyllic landscape.

In his book *Les Lacustres* describing this phenomenon, Marc-Antoine Kaeser demonstrates the power of a story when it succeeds in making the political ambitions and popular sensitivity of the moment converge. "The unprecedented and enduring attachment to these new ancestors inspires a romantic patriotism. It establishes an ideal of freedom and resistance that reconciles liberal progressivism and nostalgia for the good old days." The myth remained so ingrained in Swiss culture that proof of its fictitious nature took several decades to reestablish a more historical vision of Switzerland, even in schoolbooks.

Beyond Storytelling

When used well, developing a convincing story is an undeniable way to link an innovative proposal to users' cultural and social backgrounds. But is it enough? The most convincing examples seem to be situations where storytelling works more as a catalyst on already-fertile ground, and less as a spearhead for proposals. Nowadays, can we settle for this in work that is, by definition, at odds with what already exists? Is it enough to reduce the famous stress induced by disruptive innovation discussed in previous chapters? From the laboratory's earliest projects, it emerged as crucial to go beyond the narrative stage linked to storytelling to establish a more direct and therefore stronger link with the user. There had to be more fundamental forms of perception that gave the impression of ease, comfort, and "wellbeing" between the user and the offering – notions that are, perhaps, at the heart of

Super Normality, and could thus play a decisive role in garnering support for new proposals.

This hypothesis intuitively influenced exploration work. The enlightening reading of *Diary of a Baby* by Daniel Stern, a psychiatrist specializing in pediatrics, epitomized this, giving it an interpretation. "I was led towards a search for origins, the essence of human nature itself. We were all babies once," Stern tells us. To enable us to grasp what goes on in a baby's mind from the first few weeks until the age of four he imagines a scenario. Little Joey keeps a diary to share his vision and perception with us. For each essential stage of development, Stern completes this diary with a description of the concrete situation, followed by an analysis of what Joey recounts based on scientific research.

In the evolution of our perception, the world of stories only emerges around the age of three. It is only the fifth world to mark our evolution from birth. The first is the world of sensations. During the first weeks of our life, all the impulses experienced by our senses are thoroughly examined to create a climate, a unique landscape – a light, sound or scent. There is only a vague distinction between our inner and outer worlds, and they "may appear to belong to a single, continuous space". How we experience these stimuli depends on genetic factors and, above all, on a social construction that begins as soon as we are born. After this first world of sensations comes the immediate social world – we gradually become aware of our capacity for action, and its consequences. Simple interactions lay the foundations for social relations. The world of psychic landscapes, towards the twelfth month, plays a key role in the notion of identity. "A psychic landscape contains intentions, desires, feelings, affection, thoughts, and memories – all the events that spring up with intensity in the mind of an individual but are invisible to others." Individuals gradually discover that they can share a psychic landscape with someone else. At around twenty months a radical change occurs, profoundly altering how our thinking is structured as we enter the world of words. "The world of words is not appropriate for expressing subtle thoughts and feelings, and will sometimes lead to inner conflict," explains Stern. Nevertheless, the older worlds do not disappear. They coexist and remain present, even if the adult does not consciously pay attention to them later on.

This seeming digression regarding a baby's evolution in fact touches upon the heart of the disruptive innovation process. We have to go beyond the world of words, particularly storytelling, and directly address the earlier worlds. The designer whose mission it is to explore the uses of a new technology and incorporate it into the user's cultural and social context has to form a bond with the foundations of our sensations, plunging us back into these early landscapes and climates. He has to reconnect us to a form of "animality" that is not primitive, brutal, and bestial, but rather the result of the development of our world of sensations which, though buried, is still very much present and whose origins we now understand in social construction – a kind of primitive instinct.

This is precisely what postmodernity proposes – getting back in touch with our animality to reconnect with our roots, putting the individual back at the center of its societal project. This convergence between the perspectives of sociology and psychiatry – and the challenges of disruptive innovation – highlights the need to, once again, take this animality into account in the innovation process; we must address not only the world of words and stories, but also those buried deeper, such as the psychic world and the world of primal sensations.

This approach includes the notion of empathy. In his cult book, Victor Papanek often refers to it. The Helen Hamlyn Centre of the Royal College of Art puts it at the heart of its methods. In 2013, the Saint-Etienne International Design Biennial also gave it a higher profile by making the theme of 2013's event "Empathy or the Experience of the Other." The EPFL+ECAL Lab adds another dimension by including empathy in a wider vision. Disruptive innovation must do more than just perceive the feelings and emotions of the other in order to reflect them in a proposal; it has to integrate them into the creative processes in order to come up with new proposals that can enter into a dialog with the worlds that shape the user's perception. This involves the notions of change, encounter and exchange. As mentioned in the chapter on the digital world, the lack of such a dialog can lead to great stress for the user, as well as rejection. It can also give rise to another danger – antagonism between the proposal's promise to give the user new freedom, and the fundamentally Modernist nature of the principles underlying how it functions, which, in fact, dictates a behavior to this same individual. This contradiction does not appear explicitly but can result in visceral incomprehension, a sense of unease, with the risk of causing the user to develop a mystical drift to escape the discomfort of uncertainty and incomprehension. This drift can be associated with a pathological form of magical thinking as defined by Lucien Lévy-Bruhl. Design is not meant to be a savior, but it can contribute to reconnecting technologically useful and effective tools with the most basic nature of our culture and perception.

Interview with Nadia Bruschweiler-Stern
Disintegrated Adults

Nadia Bruschweiler-Stern, pediatrician, child psychiatrist and psychotherapist, epitomizes the transdisciplinary vision of the relationship between the child and the adult, and more specifically the relationship that shapes our personality. She campaigns tirelessly for human development to be seen as an integrated relationship between mind and the body. She also examines the ever-deepening divide between the physical and the virtual. She is author of several papers and books, and worked on *Diary of a Baby* with her late husband, Professor Daniel Stern. She currently runs the Centre Brazelton Suisse at the Clinique des Grangettes in Geneva.

Daniel Stern explains that our perception is based on several worlds that end up cohabiting and that are the result of a genetically-based social construction. But a mystery remains, despite this somewhat rational explanation. Where do we draw the line in what we bring into play with regard to design and the creative process more generally?

This gray area is fascinating. Some scientists only believe what they can measure. Other people, especially in the human sciences, observe visible results but cannot measure them. These results are often specific to the individual. We don't yet know how to grasp this field. Intuition is part of this gray area, like certain other aspects of artistic creativity. Generally, they're based on the translation from one mode to another, for example from rational data to a more sensory mode.

When exploring a new territory, the child moves away, and then comes back to his parents – towards a reassuring point of reference. Can we make an analogy here with innovation, where you have to establish cultural and social points of reference that users could always return to for reassurance after a quick foray into new features or content?

Innovation aimed at adults is not unrelated to what we can observe in small children discovering the world, but it is nonetheless different. Typically, a baby experiences development where the physical and psychic worlds are integrated. Adult, however, are no longer subject to this integrated way of functioning, and can develop their mental world in an extraordinary way, in essence detaching themselves from their senses.

Could this become an issue?

The body is slow, and our actions can't compete with the speed of thought. Thus, the body feels like an impediment, an obstacle to performance. We overdevelop our minds and disconnect from our bodies, becoming more and more disintegrated. Then we pay to rediscover these same bodies during "awareness" seminars that are so much in fashion, or to suffer at the gym on exercise bikes that don't go anywhere! We've compartmentalized our activities.

So you think it's useful to continue exploring how to give a physical dimension to digital interaction?

I think it's crucial. In his book *Forms of Vitality,* Daniel Stern also looked closely at movement, particularly the importance of its quality. How an action is performed carries a lot of meaning. It remains to be seen whether we will return to this physical dimension, or if a form of Darwinism will combine with our adaptive capacity, to push us towards an increasingly dematerialized society.

Project

Under Pressure

[1]

[2]
[3]

Soft to the touch, yet hard as a species from the tropical rainforest, this wood is mere pine. However, with its lines that echo the object's shape and a texture that can vary from rough and raw to extremely smooth, it has a unique capacity for expression. How do these radical variations come to be? The process, called densification, recalls an old dream: to truly master this material that has been used for thousands of years.

Starting in the 19th century, a greater knowledge of wood treatment and distortion techniques profoundly changed its potentials uses. In object pop culture, the Thonet Chair No. 14 – so typical of bistros and bars – illustrates this evolution as the first mass-produced chair, with its emblematic bound wooden arcs. And yet its long-term stability is limited: the ends must be constrained with judicious assembly and protected from strong variations in humidity. However, it is a process that requires no additives – just a simple combination of humidity, heat and mechanical force.
A century and a half later, a similar revolution is taking shape with densified wood. In this case, though, the transformation targeted not only form; it also changed its mechanical and physical properties. More surprisingly still, instead of shaping the wood by simply bending and binding it, densification made it possible to actually mold it, with surface detail as fine as a hundredth of a millimeter. The hardness and smoothness of the transformed material likewise offered novel machining opportunities. And, as in the process used by Thonet, densified wood contains no additives – just heat, humidity and pressure.

The idea of densifying wood is not new. The notion already existed in 1880, but the complexity of the material and its components long thwarted attempts. These difficulties were overcome either by injecting chemical substances into the wood or by focusing on only a few specific properties. For instance, high-temperature treatment ensures stability of form – even with outdoor use, and provides resistance against insects and rot. But, it destroys the wood's mechanical qualities.

In the 1990s, advances in research methods and knowledge of wood made it possible to achieve true densification. Several studies made key contributions to demonstrating its feasibility. The Parviz Navi group at the EPFL's Construction Material Laboratory was among the leaders in the field, providing stable samples of several species of wood.

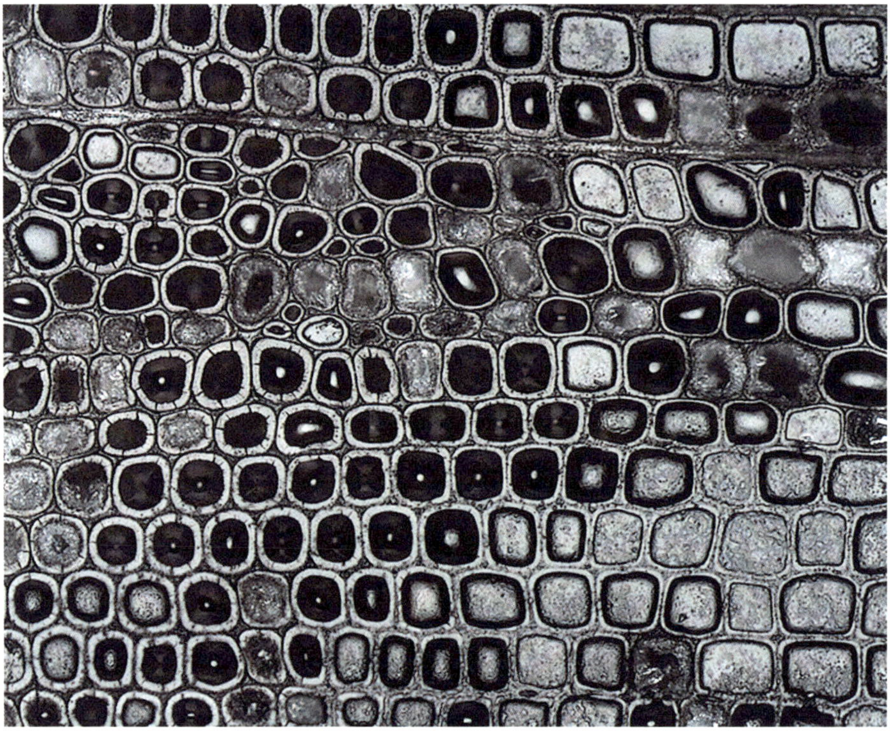

[4]
[5]

In the past decade, science made a major leap. Today, it is up to engineers to resolve important issues that will enable a move from basic research to industrialization (defining limits of size and shape, developing simulation tools capable of enhancing process control, developing efficient production lines, etc.)

However, the future of densified wood depends not only on these questions; it also requires exploration of potential uses. What new opportunities can this material offer? What formal languages can it generate? We need only think of Marcel Breuer and his steel tubing or of Verner Panton and his composites to imagine the range of possibilities when design takes a new material and explores its possibilities. In 2009 the EPFL+ECAL Lab got hold of Parviz Navi's experimental machine. Its designer, Fred Girardet, agreed to continue the adventure.

The first step was to increase the sample size which, until then, was limited to the few centimeters necessary for scientific research. As part of her degree work at ENSCI-Les Ateliers, Lea Longis, who had already participated in the Sunny Memories project, decided to take on a second challenge: to develop a formal language for this material. By creating a set of molds with simple geometric shapes, she developed more experiments, orienting the fibers, textures, visual expression and the tensions used. Substantive work brought to light an unexpected wealth of expressions, but also the delicate technical constraints of implementation. Most notably, the fibers offer great rigidity; they can impair the molding and process by creating uncontrolled shrinkage depending on the desired dimensions. Lea Longis' and Fred Girardet's collaboration provided a solid base from which to initiate the third step: to create five objects that incarnate the concrete possibilities of this material. Given the complexity and effort needed to produce one-of-a-kind experimental objects, this exploration was conducted as a series of partnerships with five renowned designers who, in exchange, were allowed to enter into the intimate world of the laboratory and to discover the secrets of this wood.

The Swiss studio Big-Game designed a door handle. The machine takes a natural pine board and turns it into a veritable three-dimensional object. Near the handle's axis of rotation, full densification combines the material's finesse and mechanical strength, while the part made for gripping has a lower densification, offering a nice, rounded feel. Normal Studio in Paris created [6] →

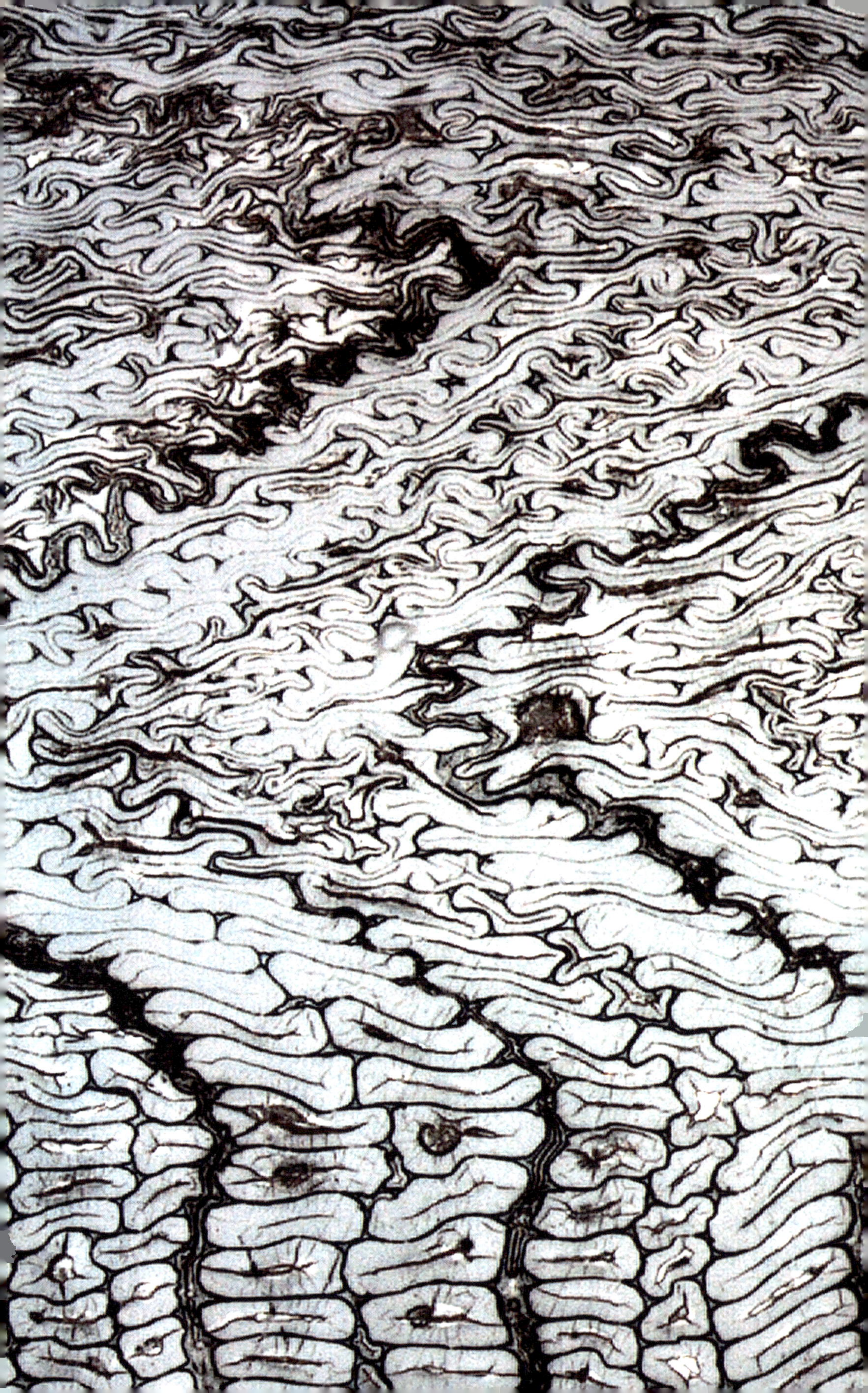

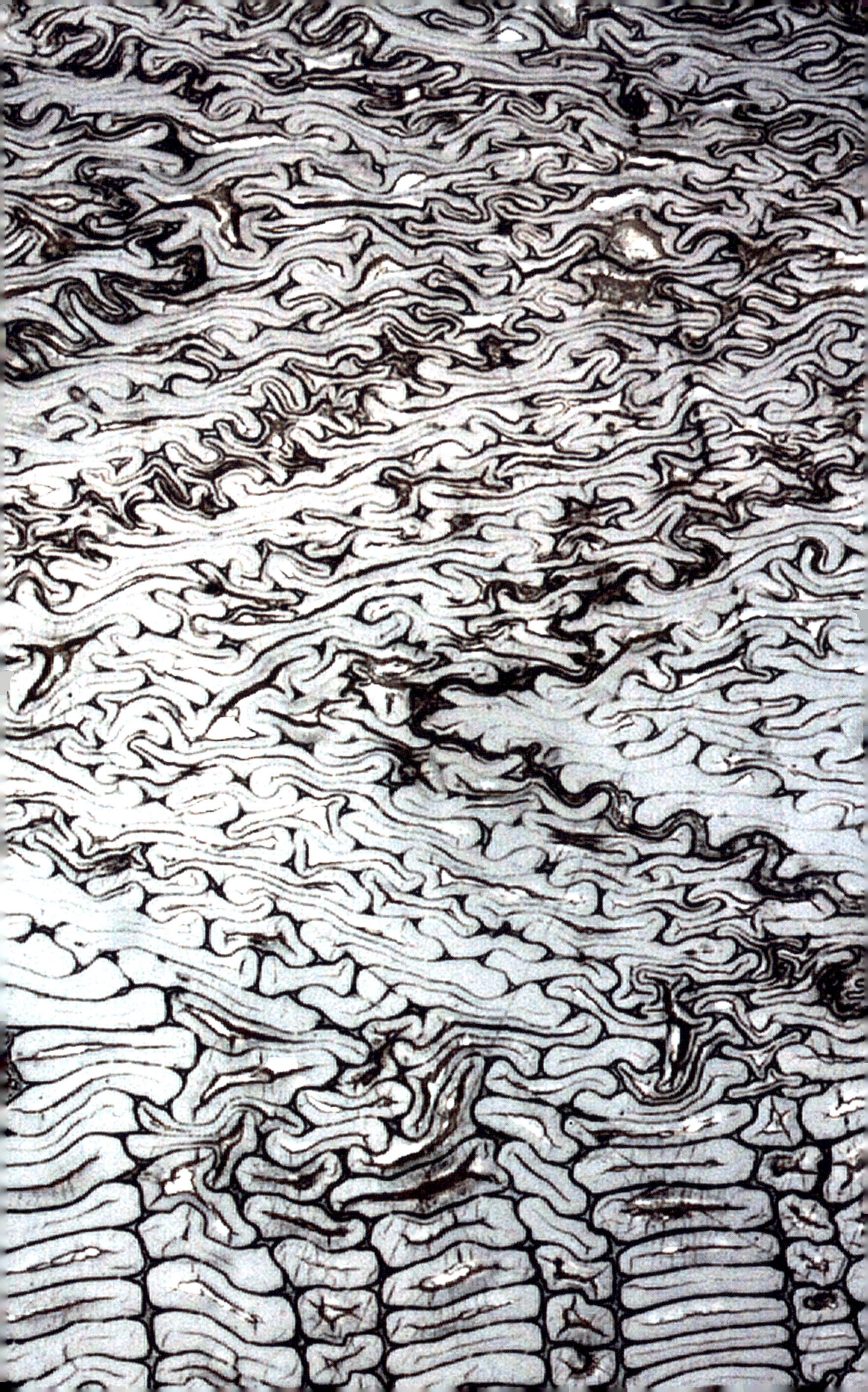

a pair of headphones in order to tackle the geometrical limits of the process, with extreme curves and subtle textures, creating a unique sense of lightness. British designer Paul Cocksedge gambled on the material's strength and expressiveness with a pair of high-heeled shoes. For the lower portion of the heel, the designer used full densification to create an ultra-slim silhouette and to make it resistant to abrasion from walking. Higher up, however, the wood underwent minimal processing, to provide a stable base and wide contact surface. The look is completed with a microscopic image of wood printed on the shoe leather. As with the heel, Cocksedge uses a lower density image for the upper part of the shoe and a completely densified one for the lower part, giving the object an organic look. Because of the quality of her work, Léa Longis also joined this group of well-established designers in creating an object. Her goal was to give a noble quality to a piece of densified pine wood by evoking iconic references to other materials, such as crystal facets. To make this idea a reality, she designed three small boxes which were inspired by romantic boxes from the late 19th century. Densified wooden lids cover Corian containers, giving the collection a contemporary identity.

Each of these projects explores prospects that go beyond technical prowess to reveal the material's capacity for expression in a variety of uses and contexts. After presenting its first three project-objects at the Design Museum, the completed Under Pressure project will make its first full appearance at the Musée des Arts Décoratifs in Paris in 2014.

CURATOR
Nicolas Henchoz, EPFL+ECAL Lab

DESIGNERS
Normal Studio: Jean-François Dingjian, Eloi Chafaï, Paris
Big Game: Grégoire Jeanmonod, Elric Petit, Augustin Scott de Martinville, Lausanne
Léa Longis, Paris
Paul Cocksedge, London
Chris Kabel, Rotterdam

ENGINEERS
Fred Girardet, Rino Sarl, Blonay
Olivier Nguyen

SCIENTIFIC PARTNERS
Prof. Parviz Navi, Bern University of Applied Sciences
Prof. Yves Weinand, Bernhard Stamm, IBOIS, EPFL, Lausanne
Laboratoire des matériaux de construction, EPFL

MICROSCOPY
Tauno Jalanti, Microscan Service SA

WITH THE SUPPORT OF
Rino SARL
Microscan Service SA
Muri AG

[1] The Normal Studio headphones play on extreme distortion to create a unique sensation of lightness and strength.

[2] The Big Game door handle has the rich look of densified wood thanks to lines that reflect the processing. The densified wood used here was originally a simple pine board.

[3] Léa Longis' boxes explore the material's textural potential. The wooden parts have not been polished – their brilliance is simply a natural effect of the processing.

[4] Paul Cocksedge's high-heeled shoes play on differences of densification between the sole of the shoe and the top of the heel. The pattern on the leather reproduces the texture of the wood microscopically.

[5] A piece of pine wood microscopically before processing. The circles show a tubular structure with large empty spaces inside.

[6] The same piece of wood microscopically after processing, at the same magnification. The densification here is complete: the tubes, now flattened, no longer show any empty space.

From the Idea to the Product

Either we can manage the innovation that presents itself to us – and there is a good chance that it is not something new – or it is something novel and therefore we do not know how to manage it. In either case, innovation keeps managers from sleeping at night," says sociologist and anthropologist Bruno Latour. After outlining proposals to overcome the apparent paradox of disruptive innovation, we must link them to the pragmatic context of the economy, production and industrialization. In other words, we must discuss how these ideas and principles born in a research setting can have a real impact on a wide audience. Without this, the leap to disruptive innovation would be lost; the target audience would not appropriate the outcome of this effort, which would never get beyond the invention phase, despite the meaning and values offered by the design.

In order to meet this challenge, the EPFL+ECAL Lab was not designed to be a typical academic laboratory. It is not part of any particular faculty, but rather is dependent on the Vice Presidency for developing and transferring the technology received from the Ecole polytechnique fédérale de Lausanne. Developing a relationship between the academic and industrial world is part of its mission, allowing it to have real ambitions in terms of innovation, through partnerships with external entities.

This chapter is based on the experience gained by the Laboratory since its inception. It does not set out to compete with textbooks or advanced training in the field of innovation. Rather, it aims to fit the proposals found in this book into a pragmatic context that could have a real impact on the production of objects and services. It also discusses role sharing between academics and industry, and proposes new types of relationships.

Some companies contact schools hoping to find a cheap design studio. To these companies, the EPFL+ECAL Lab's answer is always the same: "You're wrong on two counts. We are neither cheap, nor a good design studio." Such a discourse might seem abrupt and not very businesslike, but it comes with an explanation that clarifies the Lab's position: "If you want to develop a project whose final specifications you already know, a design studio can translate your idea into an object intended for production. However, if you want to explore how to appropriate a new technology in order to create value, and to play a pioneering role by maximizing your opportunities, our laboratory provides a suitable environment." The confusion regarding the Lab's role is understandable; innovation is trendy, but this term has a wide range of interpretations.

States, which largely fund scientific research, require technology schools to help boost innovation, especially to stimulate competitiveness at the national level. Companies also proclaim their commitment to innovation, "The key to our future growth and differentiation, vis-à-vis our competitors, is innovation, which is at the heart of our business strategy," said one telecommunications operator on one of its web pages in 2013. So there is convergence of intentions amongst politics, academics and industry. Implementation, however, is often difficult, especially in a disruptive aim, i.e. radical innovation. In 2012, Philippe Michel, project manager and head of anticipation for Orange Valley, wrote his thesis for the Innovation by Design Masters program at ENSCI-Les Ateliers in Paris, under the direction of Nicolas Henchoz. In this paper, he dissects the innovation process based specifically on the case of France Telecom, using the Minitel as an example. This small terminal, distributed to the majority of French households in the 1980s, was the first big hit of digital connectivity, before being replaced by the internet fifteen years later. Philippe Michel's analysis converges in a pertinent way with the theme of this book. Echoing Pascal Morand and Delphine Manceau's paper *Pour une nouvelle vision de l'innovation*, he confirms the need for, but also the difficulty of, uniting the approaches of technology and design: "Without a real culture on design, the company remains stuck in a dichotomous view of the discipline [design], which makes a vision focused on technology and functionality co-exist and stand in opposition to a vision centered around aesthetics and lifestyle."

He therefore suggests going beyond the traditional concept of R&D (research and development) to one of RID (Research, Innovation and Development). A key evolution, as it gives innovation a specific position, enabling it to define new cross-links with other functions, such as communication and marketing. It would be an ideal partner with which to initiate the opportunities created in an academic laboratory, to translate them into tangible products and services. Pending the emergence of such structures, several practices can forward the success of such collaboration.

The Starting Point

What technology are we talking about? Technology can simply be one specific component of a larger system and claim no special relationship with the end user. Thus, the challenge remains centered around performance, and does not require exploration through design. However, if it offers the potential for innovation – as we saw in its sociological definition – it involves raising questions of the meaning, relationships and values that can connect it to the user, to go beyond the notion of performance, thus becoming an overall experience for the user. The designer becomes the key partner for engineers, in order to explore possible practices, to make them part of users' social and cultural context, and to develop principles of expression which see these intentions to fruition. The academic dimension of this exploration is justified if the designer has the freedom to open up new fields and bring in the unexpected. The academic contribution is also relevant to understanding why an innovation does not give rise to appropriation by the public and propose new links. The wording is vague at this stage, and thus easy. Hence, it is necessary to define how to approach a project that justifies this meeting between these two very different but complementary disciplines.

Which Strategy?

There is no magic recipe. Each project requires a specific and dedicated approach, on the one hand, because the basis of each issue is different. Sometimes the challenge lies in creating opportunities for a very specific technology. Sometimes it is a case of misunderstanding or disinterest that, over time, undermines our relationship to a particular type of device or service, as evidenced in the case of the remote control with the Lazy Bytes project. Certain topics, such as the solar cells of Sunny Memories, involve cutting-edge science, but utilize the basic components of design: shape, color, and pattern. Calling on a multitude of design students from different schools provides a flowering of ideas and increases the chances of successfully improving our relationship with solar energy. Other challenges, still, require a great deal of closely-interlinked, highly-specialized knowledge, such as augmented reality, which brings together the physical and virtual worlds, objects and animation, and technological principles and digital tools. A field too vast to manage for the future designer who is just learning the basics of his craft. The complexity and cost of prototyping is also an important criterion in developing a strategy. Testing the value of a proposal requires

conducting a credible, representative demonstration for the user. No one can actually predict user perception, which combines a complex mix of cultural, societal and sociological factors. These tests require the creation of models (often functional) capable of generating an accurate experience. This can sometimes mean significant production efforts and deadlines, which are essential to the seriousness of the outcome. The project Under Pressure, for example, required extremely precise milled aluminum molds and numerous trials – hardly compatible with the freedom of trial and error often given to students, which also resulted in increased attempts.
Finally, a project strategy must take into account partner-related constraints, such as privacy. A class must be able to share its ideas as openly as possible. One key parameter remains in regard to strategy – time and scheduling.

For How Long?

We have seen that issues of time are at the heart of the relationship between engineering and design. The birth of a disruptive technology in an academic process generally requires many years between the first hypothesis and final scientific demonstration. Often, it emerges elsewhere within the same laboratory, during a process of fundamental questioning, far from industry constraints. Then comes the next step: transforming a laboratory demonstration into an industrial product that is reliable and competitively priced.
For disruptive innovation, exploration through design must take place within a specified period between these two stages: early enough to maintain its freedom of exploration, but late enough to test ideas in a concrete form. While the creation process in the design phase is faster than that in engineering research, it usually requires cycles of six to eighteen months from the start of the project to delivery of a series of ideas in the form of convincing models, to then be followed by test and communication phases. Real products may come out of this phase, but the major benefit comes primarily from the knowledge generated, which can affect strategic decisions and entire product lines, as well as increase a company's competitiveness and positioning. The mirror of the Tattooar project is emblematic of these two facets: augmented reality applied directly to the arm or wrist of a user is actually a fun and immersive installation in an event context. But its value primarily has to do with the knowledge

it generates for jewelry. For instance, it can provide important information to enhance precious, durable objects with a message which evolves over time. It is therefore critical that the duration of the project be accompanied by a schedule that takes into account the specific characteristics and needs of each discipline in order to utilize them effectively.

How to Convince?

The event context and public presentations also add significant value. The public, often overwhelmed by the hustle and bustle of everyday life, rarely has sufficient time to appropriate radically new proposals. Despite efforts to create relevant usage scenarios and to establish cultural and emotional links that give the proposal legitimacy and meaning, the user may still miss out on these proposals – particularly given that disruptive innovation does not necessarily respond to existing demand, but rather offers new possibilities. Exploration through design therefore has a further mission: to take advantage of the expressive quality of its proposals in order to sensitize the target audience to the prospects that will soon be available to it. The Lazy Bytes trials clearly demonstrate that, given the choice, people prefer remote controls similar to the general shape they are familiar with. This process of sensitization is similar to current practices in the automotive industry, which presents concept cars at every fair. In the IT industry, it often appears in the form of information leakage to major media. But it more rarely applies to disruptive innovation, which cultivates the sharing of information within confined academic communities and the respect of intellectual property.

Who Does What?

The context in which the various facets of disruptive innovation are practiced plays an important role. Every business school has a culture, dimension, and organization that are specific to it. The division of tasks plays a vital role in the success of a project. In general, the academic world has this famous freedom that allows it to elicit surprise and to take advantage of the unexpected, without compromising the rigor of its approach. The industrial world has knowledge of the market and its constraints, as well as skills specific to industrialization (reliability, optimization, production, distribution, etc.) So commonplace, this principle nevertheless proves difficult to implement. Initial clarification of the roles makes it possible to focus the relationship on this complementarity, like the Lazy Bytes project, conducted with the Kudelski Group. From the beginning of the project, Thierry Dagaeff, head of innovation for digital television, presented the students with the challenges of the market, but without restricting their creativity. The students and schools thus were entitled to their freedom – but not to ignorance. EPFL+ECAL Lab not only launched the theme and set up the project, but was also responsible for translating the designers' intent expressed into the language of the industry and extracting the overall knowledge, relevant for both industry and the academic world.

Royalties or Flat Rate?

The vast majority of schools establish their relationships with manufacturers based on a similar principle: the partner pays an initial sum for the project and has a priority right to then negotiate the intellectual property rights on the outcomes it considers most promising. The principle seems perfectly fair and logical. EPFL nevertheless questioned it. The result: more resources for the school and greater satisfaction for industrial partners! The explanation here lies in bringing these principles face to face with the reality in the field. When a partner identifies a relevant result at the end of the project and negotiates with a school, the assessments and perceptions differ. EPFL optimistically imagines the benefits the industry will reap from the work it has done, while the manufacturer above all sees the efforts still necessary to turn the idea into an actual product line, the difficulties linked to industrialization and stadardization competitiveness issues, risks, etc. The dialogue becomes even more complex when it comes to assessing the extent of the laboratory's contribution to the finished product, which often changes considerably before finally arriving on the market. Lastly, how to control manufacturers' sales figures in order to ensure the accuracy of the royalties? All the more so since, for some companies, these figures are of strategic importance and therefore must remain confidential. These practical difficulties often lead to a stalemate in the negotiations, while a key factor for success is the speed of action in term of reactivity, in the form of the famous "time to market." After a thorough analysis of the service of industrial relations, EPFL changed its policy: when a project begins as part of an industrial partnership, it requires a higher initial payment, but also includes the right for the manufacturer to use the results that are of interest to it – a right generally limited to the company's area of activity. It also carries with it specific conditions, including a licensing principle that guarantees researchers the right to continue their research on the subject unhindered, and to publish their results in scientific journals after a sufficient period, to ensure their ownership of the intellectual property rights (usually after a several months).

Back, then, to a central question: is it ethical for a manufacturer to make huge profits on the results of a student or assistant without paying substantial additional sums? There are two possible answers to this. The first concerns the number of actual cases, which are very rare or nonexistent for most schools. The second answer is a choice: should we not foster the resources allocated to the project as a whole, rather than protect the very hypothetical interests of a specific part? The answer may differ depending on the context, of course.

The initial commitment of industrial partners in most of the projects presented here have achieved new breadth and diversity, with the participation of up to a hundred students and teachers from the four schools during a semester. It has also provided the means to realize complex ideas and bring them face to face with a target audience, thus generating new knowledge. Finally, it has supported the creation of exhibitions and installations that have been exposed at prestigious galleries and festivals in Paris, London, Kyoto, Berlin, New York and San Francisco and, hence, given international visibility to the many students selected.

The academic world requires substantial resources to meet the challenges imposed on it, but it must also gain perspective in order to develop effective relationships with its partners, to produce a genuine, valuable impact to the greatest number – to its students and its researchers first, who produce this knowledge, but also to society as a whole by realizing new ideas, new services and new activities.

Interview with Yves Béhar
Reinventing the Design Business

Yves Béhar is one of the most influential designers of his generation, but is more likely to send his friends pictures of surfing than of design! Much more than just a personal hobby, his love of surfing is closely linked to his work. His products tell a story, evoke a connection, but above all arouse sensations and emotions. Being able to catch a wave and glide on the water with a surfboard involves an almost fusional relationship between the body, the object and the environment; it is a physical and mental experience in which skills and intuition meet. This strength can be found in his creations, which often use technology to enhance the user's experience. One Laptop Per Child, a computer that aims to offer every child access to information networks, even in the most underprivileged areas, combines electronic and energy challenges with those of communications and materials. The Leaf Lamp played a pioneering role in LED lighting. Yves Béhar creates both unique products, like his monumental creations for the Swarovski Crystal Palace, and mass-produced ones, such as Jawbone earpieces. His approach is a reference for the EPFL+ECAL Lab, where he was also involved in the Sunny Memories project. Born in Switzerland, he continued his design studies at the Swiss branch of Pasadena's Art Center College of Design, and then emigrated to the United States. In 1999 he founded his company, Fuseproject, based in San Francisco.

You often use new technologies in your products. What kind of relationship do you have with these technologies?

They add a new dimension to experience. In the 1950s and 60s, when designers like Dieter Rams clarified the question of the product's function, there was already something magical. This marked the beginning of "User Interface Design." The interface design between users and technological devices comes from industrial designers. But a new dimension opened up during my studies at the beginning of the 1990s – using technology as a tool to create a new product-based experience. The Learning Shoe that I proposed at that time prefigured the "internet of things." With its data chip located in the sole, the shoe recorded users' behavior in order to propose a model that was entirely adapted to their use the next time around. Thus, it proposed a new relationship between the object and the user, and between the user and the brand. The shoe grew closer to the physical and emotional needs of each person who wore it.

A lot of designers appear to be either fascinated or repelled by technology. Yet, it seems to be a completely natural part of your designs...

This is the difference between a designer approach, which focuses on the human experience, and technophiles, who are more interested in technology's capacities. We're seeing a lot of products that take features too far. A designer should use technologies that make a difference in everyday life, but also be able to reject unnecessary features.

In your work, these technologies often serve a sensual purpose, like the materials for the Philou soap packaging, or the sensor of the Leaf Lamp.

Only an emotional and sensual approach can give the object such strong potential for communicating its purpose and the ideas it conveys. I see the emotional dimension as an essential part of how it functions.

With such a focus, your creations actually often go beyond storytelling...

For me, the notion of storytelling is a personal commitment – a "one-on-one" with the user – and is therefore the best way to understand our intentions and ideas about one brand versus another. This commitment is translated through form, texture, and the density of the materials. Design is the first expression of human intention. It is part of the relationship between the intention, talking about it and making it. Design is not marketing or advertising work, but it does impact them.

How do you actually translate this into a concrete relationship with a company?

I suggest working together in partnerships that go beyond the usual notion of a design commission. From the very beginning, I wanted to make lasting work – things that would still be important as time went on. The partnership allows you to share and construct visions and objectives linked to a brand. In a long-term perspective like this, the design of objects improves as the relationship grows.

The idea is obviously appealing, but it must not always be easy to convey this to companies, who have other obligations for their business model.

But design is hard work! It takes years to build that kind of relationship, and it doesn't work with every company. It also means reinventing the design business itself, which needs to rediscover the value of the work of Dieter Rams, Charles and Ray Eames, and George Nelson. They made major contributions, and their works have endured the tests of time.

Eames and Nelson helped build Herman Miller's reputation. Your Leaf Lamp sealed the beginning of your partnership with this company. It's an example of disruptive innovation that's now in New York's Museum of Modern Art.

The Leaf Lamp really started with the concept of the user's experience. I always wondered why light was adjusted according to intensity, not tone – especially cold light, which is good for working, versus warm light, which creates a pleasant atmosphere.
The arrival of LED technology offered the possibility of switching from one tone to another. At the same time we worked on the control system, with a unique sensor to give it a new physical feel. Herman Miller followed the developments, but we had to prove what these ideas – combined in a single lamp – could produce when it came to experience. We formed our own in-house technical team to resolve various issues, and set up a demonstrator system. The project took five years in all, but in the end we managed to put the first LED lamp for the general public on the market.

Project

Seeds of Colors

[1]

Edson Matsuo is the creative director of Melissa, the flagship brand of a Brazilian company specializing in plastic shoes. He's one of those people you never forget. Behind his glasses that bring to mind Le Corbusier, his eyes twinkle with vivacity and ideas abound at lightning speed. Edson Matsuo is the very personification of Melissa's flourishing identity. "I love plastic!" he enjoys saying. Through his choices, he is redefining the material's expressiveness by playing with its colors, flexibility and virtually infinite forms, to the point of making Melissa a symbolic brand throughout Latin America. The company works with renowned Brazilian designers like the Campana Brothers, as well as designers from elsewhere such as Vivienne Westwood, Gaetano Pesce and Karl Lagerfeld. It is hoping to conquer other continents as well, and has opened a showroom in New York's Soho district that, like the Sao Paulo showroom, aims to offer clients not just shoes, but a new experience. Edson and his teams worked on the ambiance of this space, with the delicate task of transposing Melissa's identity to the more codified, sophisticated world of New York. The elongated space stands out with its bright atmosphere thanks to very pale walls structured by geometrical shapes. Despite a strong identity, the space draws attention to the shoes themselves. The back of the showroom serves as a projection area. Just in front of this wall is a circular hole in the ground, a few meters in diameter, creating a kind of well, or cave, as Edson Matsuo describes it. A staircase invites visitors to go down to see special items, such as new models. Known as Galeria, this showroom has a strong cultural dimension and hosts artistic events. Under the leadership of Laetitia Wolff, a journalist and project curator in New York, Edson Matsuo has put forth a new challenge to the EPFL+ECAL Lab – to transform the overall perception of the space, and more specifically to enhance the role of the cavern. The challenge gave the Lab the perfect opportunity to apply and experiment with research on mixed and augmented environments, led by the designer Thomas Eberwein. The challenge was no longer to create an installation in a sales outlet, but rather to offer an overall experience that unites the physical space with the colorful world of Melissa.

Several avenues were explored for this project. The first focused on the perception of space, giving it a clear hierarchy. The cave had to play a central role, from which Melissa's cheerful energy would

erupt. The projection area at the back of the showroom would be a result of this source, thus becoming an extension of the showroom's main area, a place to contemplate, choose, try on and discuss – still in Melissa's image, but without being disturbing.
The second avenue was to develop a kind of interaction that was not only playful but also legitimate – a far cry from the simple virtual effects of aquariums or dead leaves in the entrances of restaurants or shops lacking ambiance. The idea of interacting with the shoes down in the cave seemed an obvious choice to trigger visual events that would create a link with the main floor and influence its atmosphere. The third line of investigation involved the visual language. The range of bright colors used by Melissa for its collection seemed essential, but in what form? Pure abstraction? Symbolic expressions? Explicit representations? The answers to these questions had to take into account the constraints of space and utilization, as well as technical limitations such as the resolution of the projected images.

Working closely with the engineer Daniel Tamburrino, Thomas Eberwein devised and developed a scenario: descending into the cave, visitors arrive in the world that visually symbolizes Melissa's energy and creativity. By projecting an image onto the floor, visitors walk through a space where seeds of color twirl about. The ground seems to open up under their feet, causing the flurry to appear even deeper within the cave. Visitors find themselves in the center of an interaction that affects the whole environment; their feet attract the dancing particles, which then transform into streaks of light. These streaks shoot out towards the walls, climbing the large projection wall to become tall grass. Gradually a forest of tall, colorful grasses becomes denser as more visitors arrive. The grass ripples gently, naturally, like a virtual garden in the extension of the main showroom area. After an initial trial phase, the EPFL+ECAL Lab asked designer and architect Michael Szivos and his team from SoftLab to complete the physical dimension of the organization. Most notable, it was important to capitalize on the space's height with an aerial presence. Szivos responded with a proposal of clouds. Plastic films create an envelope around blown-up balloons, echoing the material, color and idea of particles. Near the entrance, a Thomas Eberwein video projected on a giant LED screen presents the installation from the depths of the cave to the tall grasses. This video creates a visual climate

[2] →

upon entering the store, and at the same time immediately gives visitors an idea of what they are about to experience. Finally, a sound design related to the events in the cave adds the final touch. Inaugurated in July 2013, this project went beyond the scientific definition of augmented reality based on computer vision, entering a veritable dimension of expression and perception.

Seeds of Colors does not show the performance of computers, projectors or detection systems, but rather creates an overall experience that "augments" the space with the Melissa's identity. The color and movement, the allusion to a world of abundant vegetation, the symbolic play between the forces of the earth and the lightness of the living evoke this joyful world in which every step is a source of creation.

CURATOR
Nicolas Henchoz

DESIGNER
Thomas Eberwein

ENGINEER
Daniel Tamburrino

SCENOGRAPHY
Michael Szivos, SoftLab

SOUND DESIGN
Dave Meckin

NYC CONSULTANT
Laetitia Wolff, Futureflair

[1] Seeds of Colors tangibly applies principles coming from an EPFL+ECAL Lab research project led by Thomas Eberwein. The physical space, the digital content and the brand universe constitute a whole experience where the product plays the central role.
Photo Thomas Eberwein.

[2] In the upper section of the installation, large reeds of colorful grass grow into the showroom with organic swaying movements – a natural extension of the living space.
Photo Thomas Eberwein.

And Now…?

From Joey, who is only a few weeks old, to Betty, surfing the web at 90, this book is about the meeting between design and engineering, as well as the relationship between life and innovation. It plunges into the adventure of a new laboratory to extract ideas, strategies and findings and, more importantly, to consider them relative to disruptive innovation, design trends, and changes in society. This book attempts to offer new insight into how to unite these two fields at the heart of this form of innovation, and to create an approach to facilitate their future based on the research conducted by the Lab. Historical references play an essential role in understanding current issues and exploring new models, even within a context of rupture. Although the themes have certainly changed – especially with the emergence of digital technology – the contribution of each discipline is part of a continuum. Each of the projects done at the Lab that we have described here contributes to the ongoing construction of this approach, like white stones marking the way.

In Lausanne, this path continues to develop and expand. EPFL and ECAL are entering a new phase of collaboration, starting with the creation of a joint postgraduate studies program. Design and engineering graduates will work side by side, each maintaining their own specificity and approach. This program will take participants to the heart of real innovation projects, increasing their capacity for research and exploration. It will also build on lessons learned at the bachelors and masters levels, and raise awareness of the new challenges linked to innovation, giving students the tools with which to tackle transdisciplinary projects combining different types of knowledge, especially design and engineering.

Alexis Georgacopoulos took over as head of the ECAL from Pierre Keller in 2011, in the spirit of continuity and with the same key ideas – rapidity, intuition, relevancy, and novelty. And one more: "not *taking* oneself seriously, but *doing* things seriously." During the Keller era, Georgacopoulos was head of the industrial and product design department. In his new role Georgacopoulos taken on new challenges, especially as regards digital technology. The EPFL+ECAL Lab will play a pivotal role in maintaining a direct link with technologies emerging from the research laboratories – a strategic decision consistent with the EPFL's evolution which, in turn, designs training to make emerging digital tools available to the social sciences, humanities and the medical field.

The Lab will also continue developing projects that address the societal challenges linked to new technologies, such as the relationship between the physical and digital worlds, and the paradox between the use of digital tools and the societal rules they give rise to. Some of the Lab's industrial partners have become aware of these issues and are gearing up to become pioneers. Faveeo, a start-up created by entrepreneur Alexis Dufresne, strives to open up new horizons in the quest for information and knowledge organization. It has also taken a firm position with regard to transparent Internet browsing: let users see the criteria the machine uses, choose content corresponding to their preferences and, more importantly, to allow them to *modify* these criteria later on. Dufresne's sociology background undoubtedly allows him to play a special role in this area. Early partners in the Lausanne lab's creation, like Pierre Keller (former ECAL director) and Christophe Guignard (a lecturer at ECAL) – founders of the company Fabric|ch – have strengthened their commitment to the representation and materialization of virtual environments like the cloud, that intangible space where data exists, so that it remains accessible at all times. Their research is part of a tradition at the interface between sociology, physical space and the digital world, like the concepts of "radicant" and "creolization," which address the blending of individuals' knowledge and mobility in the 21st century.

Yet, this emphasis on digital technology does not overshadow advances in terms of materials, where major research is always on the horizon. As we saw in Under Pressure, wood cannot only be densified but also welded and assembled, without any additives, through a sophisticated geometric interlocking principle. Glass is also on the brink of a major revolution, thanks to a new means of mastering light, and is likely to permanently change our relationship to the container and its contents. We could go on listing the myriad paths currently opening in so many fields – from the environment to energy. But one in particular bears a special dimension: knowledge of the brain and, hence, phenomena of perception. This knowledge opens up an important field of research, making disruptive innovation a conduit of long-term prospects and strengthening its dialogue with other disciplines. Such a dialogue will require rigorous project management: combining two disciplines means creating a single transdisciplinary relationship. Combining ten, however, creates forty-five relationships to manage among highly specific fields! The prerequisites for this (uniting differences, maintaining the freedom of exploration so essential to academic culture, and adopting a pragmatic outlook allowing for concrete testing of results) have already been laid out earlier in this book. As such, we should constantly be aware of the approach's overall aim: to leave sufficient room for intuition, surprise and the unexpected in the dialogue with rational thought. This adventure will therefore not be just a tidy inventory of proposals and projects, but an increasingly vast and fertile field, where disruptive ideas capable of feeding a society hungry for meaning, amidst an onslaught of technological changes, grow.

Appendix

Endnotes

1 The Industrial Revolution is characterized by the transition from a predominantly agricultural and artisan society to a commercial and industrial one, whose ideology is technical and rationalist, starting in the 1840s.

2 Popularized by German philosopher Friedrich Engels and British economic historian Arnold Toynbee, it refers to the historical process theorized during the 19th century that causes to shift – more or less quickly depending on the country and region – predominantly agrarian, artisan societies to commercial, industrial societies.

3 Literally Arts and Crafts.

4 A typical example of this is the entrances to Parisian Metro stations, designed by Hector Guimard in 1900.

5 In his article Ornement et Crime (1908), Adolf Loos wrote (about ornamentation): I have emerged victorious from a 30-year battle. I have freed humanity of superfluous ornamentation. "Ornament" was once a way of saying "beautiful"; today, thanks to my entire life's work, it is a way of saying "of inferior worth." Obviously, the echo that comes back thinks it is voice itself... I know that humanity will thank me one day, when the tramps saved will be beneficial to those who, until now, have been excluded from the goods of this world."

6 Dieter Rams was head of design at Braun from 1961 to 1995.

7 Gestalt Theory is the psychology of form, or Gestaltism (from the German Gestaltpsychologie). It theorizes that the processes of perception and mental representation spontaneously treat phenomena as forms, and not just as a simple addition or juxtaposition of elements.

8 Journalist, architecture, and design critic, and curator of exhibitions, Gilles de Bure was likewise commissioner of numerous exhibitions, director of the Galerie d'actualité at the Pompidou Center at the end of the 1970s and the first director of the Grande Halle at the Villette in the 1980s. He died in July 2013, as we were writing this book. We would like to acknowledge his critical thinking and his contribution to the field.

9 Roger Tallon is a key figure in design in France. His career is emblematic of French style design and is characterized by its domestic implications in its early stages (Téléavia portable TVs, etv.), and later its institutional and artistic implications before becoming more strictly industrial (TGV, Mexico City Metro, Montmartre Funicular in Paris, etc.). According to Tallon Gilles de Bure and Chloé Braunstein-Kriegel in their book.

10 It is important to mention here Archizoom's No-Stop City project. In 1969, in conjunction with its experimental work in the field of design, the Archizoom group conducted research on the city, the environment and mass culture, resulting in the No-Stop City project. This work revealed a "city without end," combining objects, consumer society and architecture. No-Stop City is a city without qualities, in which the individual can develop his or her habitat in a creative, free and personal way.

11 Adopted at the First World Congress of Transdisciplinarity, Convento da Arrábida, Portugal (Novembre 2–6, 1994).

Bibliography

ALTER, Norbert, *L'innovation ordinaire*, Puf, 2000.

BACHELARD, Gaston, *La formation de l'esprit scientifique*, Vrin, 1971.

BENYUS, Janine, *Biomimicry : Innovation Inspired by Nature*. Perennial, 1997.

BOURRIAUD, Nicolas, *Radicant*, Denoël, 2009.

BRIN, Sergey, PAGE, Larry, *The PageRank Citation Ranking : Bringing Order to the Web*, Technical Report, Stanford, 1998.

CLODONG, Olivier, CHÉTOCHINE, Georges, *Le storytelling en action*. Editions Eyrolles, 2010.

COLE, Henry, *Fifty Years of Public Work of Sir Henry Cole*, K. C. B., Accounted for in *His Deeds, Speeches and Writings*, Henry Cole, Alan Summerly Cole, Nabu Press, 2013

CORNWELL, Benjamin, LAUMANN, Edward O., SCHUMM L., Philip, University of Chicago, The Social Connectedness of Older Adults : A National Profile. *American Sociological Review*, 2008, VOL. 73.

CZAJA, J. Sara, Joseph Sharit, Sankaran Nair & Mark Rubert, Understanding sources of user variability in computer-based data entry performance, *Behaviour & information technology*, 17, 282-293, 1998.

DESCARTES, René, *Discours de la méthode*, 1637.

DORNER, Peter, *Le Design depuis 1945*, Thames & Hudson, 1993.

GARGIANI, Roberto, *Archizoom Associati 1966-1974, de la vague pop à la surface neutre*, Electa, 2007.

GUIDOT, Raymond, *Histoire du design de 1940 à nos jours*, Hazan, 2004.

GRAVES, Guillian, *Biomimesis*, ENSCI-Les Ateliers, 2012.

KAESER, Marc-Antoine, *Les lacustres*, Presses polytechniques et universitaires romandes, Collection Le Savoir suisse, 2004.

KUHN, Thomas. S., *La structure des révolutions scientifiques*, Champs, 2008.

de LAMARTINE, Alphonse, *Harmonies poétiques et religieuses*, Hachette, 1901.

LATOUR Bruno, *L'impossible métier de l'innovation technique*, in MUSTAR, Philippe and PENAN Hervé, *Encyclopédie de l'innovation*, Economica, Paris 2003.

PENAN, Hervé H., *L'Encyclopédie de l'innovation*, Economica, 2003.

LIPOVETSKY, Gilles et SERROY, *Jean, L'esthétisation du monde,* Gallimard, Paris 2013.

LIPOVETSKY, Gilles, *Les temps hyper-modernes*, Grasset, 2004.

LYAUTARD, Jean-François, *La condition Post-moderne. Rapport sur le savoir*, Ed. de Minuit, 1979.

MAFFESOLI, Michel, *Au Creux des apparences, pour une éthique de l'esthétique*, (1990) La Table Ronde, Paris, 2007.

MAFFESOLI, Michel, *Eloge de la raison sensible*, (1996) La Table Ronde, Paris, 2005.

MAFFESOLI, Michel, *Le Réenchantement du Monde,* La Table Ronde, Paris, 2007.

MAFFESOLI, Michel, *Le Temps des tribus, le déclin de l'individualisme dans les sociétés de masse*, (1988), La Table Ronde, Paris, 2000.

MAFFESOLI, Michel, *Le temps revient. Formes élémentaires de la postmodernité*, Desclée de Brouwer, Paris, 2010.

MARCHAND, Victoria, *Zanox The Meetic of B2B*, Cominmag, 2013.

MELO, Michka, *Le biomimétisme nous mène-t-il au développement durable ?*, EPFL, 2012.

MELO, Michka, GRAVES, Guillian, *Versus*, ENSCI-Les Ateliers, EPFL, 2012.

MICHEL, Philippe, *Du devoir d'innover au pouvoir innover*, ENSCI-Les Ateliers, 2012.

MIDAL, Alexandra, *Design, Introduction à l'histoire d'une discipline*, Agora Pocket, 2009.

MINTZBERG, Henry, *Le management. Voyage au centre des organisations*, Editions d'Organisation, 2004.

MORAND, Pascal, MANCEAU, Delphine, *Pour une nouvelle vision de l'innovation*, La Documentation française, 2009.

MORRISON, Jasper, FUKASAWA, Naoto, *Super Normal, the Sensation of the Ordinary*, Lars Mueller, 2007.

Musée international de la Croix-Rouge et du Croissant-Rouge, *L'Aventure humanitaire*, Infolio, 2013.

NAP, Henk Herman, *Stress in Senior Computer Interaction*, Thesis, Eindhoven University of Technology.

NAVI, Parviz, SANDBERG, Dick, *Thermo-Hydro-Mechanical Processing of Wood*, EPFL Press 2012.

PAPANEK, Victor, *Design for a real world*, Van Nostrand Reinhold, 1984.

ROUSSEAU, Jean-Jacques, *Du contrat social*, chez Marc-Michel Rey, Amsterdam, 1762.

SALMON, Christian, *Storytelling : La machine à fabriquer des histoires et à formater les esprits*, La Découverte, 2007.

SCHELLING, Hans-Rudolf, SEIFERT, Alexander, *Internet Nutzung im Alter – Gründe der (Nicht-) Nutzung von Informations und Kommunikations-technologien (IKT) durch Menschen ab 65 Jahren in der Schweiz*, Université de Zurich, 2010.

SILVERSTONE, Roger, HIRSCH, Eric, *Consuming Technologies : Media and Information in Domestic Spaces*, Routledge, 1992.

STERN, Daniel, *Le journal d'un bébé*, Editions Odile Jacob, 2004.

ZIEGLER, Peter, *You, Me and Everyone We Know*, Rapport interne, Helen Hamlyn Center, 2013.

Special Thanks

This work results from many years of reflexion, creation, projects, exchanges and debates with very numerous people. It is thus the fruit of a collaborative work.
Both authors wish to thank:

Chloé Braunstein-Kriegel and Jean François Dingjian for their expertise in the review of the manuscript.

Olivier Babel, director of Presses polytechniques et universitaires romandes for his commitment and attentive reading.

Michel Maffesoli, Andrea Branzi, Jasper Morrison, Nadia Bruschweiler-Stern and Yves Béhar for their availability and their contributions essential to the content of this book.

Jessica Strelec for the English translation of this work and also for her critical vision on the whole project.

Adrienne Corboud-Fumagalli, Vice-President of the Ecole polytechnique fédérale of Lausanne for her vision of the innovation and the place granted to design.

Patrick Aebischer and Pierre Keller, without whom this adventure would never have been possible.

The EPFL+ECAL team as well as all the artists, designers, engineers and scientists who contributed to the projects described in this book. With a particular mention to Alessandra Rovero Michel and Anne Viscolo for their management abilities in an interdisciplinary environment often complex.

Inka, Axel and Victor for their patience, their joy and their energy so communicative.

Jean-Pierre, Matthieu, Bernard and Jeanine for their moral support and their permanent encouragement.

Finally, the authors thank quite particularly the Fondation ECAL+ as well as the Fondation Casino Barrière Montreux.

Nicolas Henchoz

Nicolas Henchoz is director of the EPFL+ECAL Lab. Born in 1967, he studied material engineering at the Ecole Polytechnique Fédérale de Lausanne, and then journalism. He worked as a reporter for the *Journal de Genève*, and then at Télévision Suisse Romande, where he was also a newscaster. His independent work brought him to the position of art director before joining the EPFL as a member of the Presidency. He founded the EPFL+ECAL Lab in 2007. The curator of numerous exhibitions at venues such as the American Institute of Architecture in New York, the Royal College of Art in London, The Laboratory@Harvard or the Lieu du Design in Paris, as well as author of several articles and lectures, he was named *Chevalier des arts et des lettres* by the French Minstry of Culture and Communication in 2011.

Yves Mirande

Yves Mirande is a journalist specializing in design. He has degrees in both the applied arts (ESMOD) and in sociology (Master 2 from the Sorbonne), and is currently working on his PhD at the EHESS. He is in charge of design pages for the magazine *Numéro* and also works with the magazines *L'Express Styles, Intramuros, Madame Figaro, Maison Française, Les Echos Série Limitée*. He never ceases to showcase new talented designers and the most prospective projects, and is constantly analyzing social emergences in the field of applied arts and design more specifically. He likewise expresses himself through his work as a curator. His exhibitions include *Futur Archaïque (Passage de Retz, Paris, Grand Hornu, Belgium and MUDAC, Lausanne, Switzerland) and Mouvement (Carreau du Temple, Paris). He is co-author of the book Qu'est ce que le design aujourd'hui*, published in October 2009 (Beaux Arts Editions). Since 2010, he is also a consultant for major international luxury brands.

The publisher and the authors express their thank to the Ecole polytechnique fédérale de Lausanne, the Fondation ECAL+, and the Fondation Casino Barrière Montreux for their generous support towards the publication of this book.

Photographs: Tonatiuh Ambrosetti, Daniela Droz, Olivier Pasqual, Joël Tettamanti, Christophe Fillioux, Michel Bonvin, Alan Tansey, Alain Germond.

Graphic design and production:
Robert Huber, Lausanne

Photolithography and printing:
Stämpfli Publications SA, Berne

The EPFL Press is the English-language imprint of the Fondation des Presses polytechniques et universitaires romandes. PPUR publishes works of teaching and research of the Ecole polytechnique fédérale de Lausanne and other universities and institutions of higher education.

www.epflpress.org

First Edition

ISBN 978-2-940222-74-2 (EPFL Press)
ISBN 978-1-138-78278-5 (Routledge)

Routledge
Taylor & Francis Group

Taylor & Francis Group Ltd
2 Park Square, Milton Park

Abingdon, Oxford, OX14 4RN, UK
Routledge is an imprint
of Taylor & Francis Group,
An informa business.

Simultaneously published in the USA
and Canada by Routledge,
711 Third Avenue, New York 10017
www.routledge.com

Library of Congress
Cataloging-in-Publication Data
A catalog for this book is available
from the Library of Congress.

Printed in Switzerland

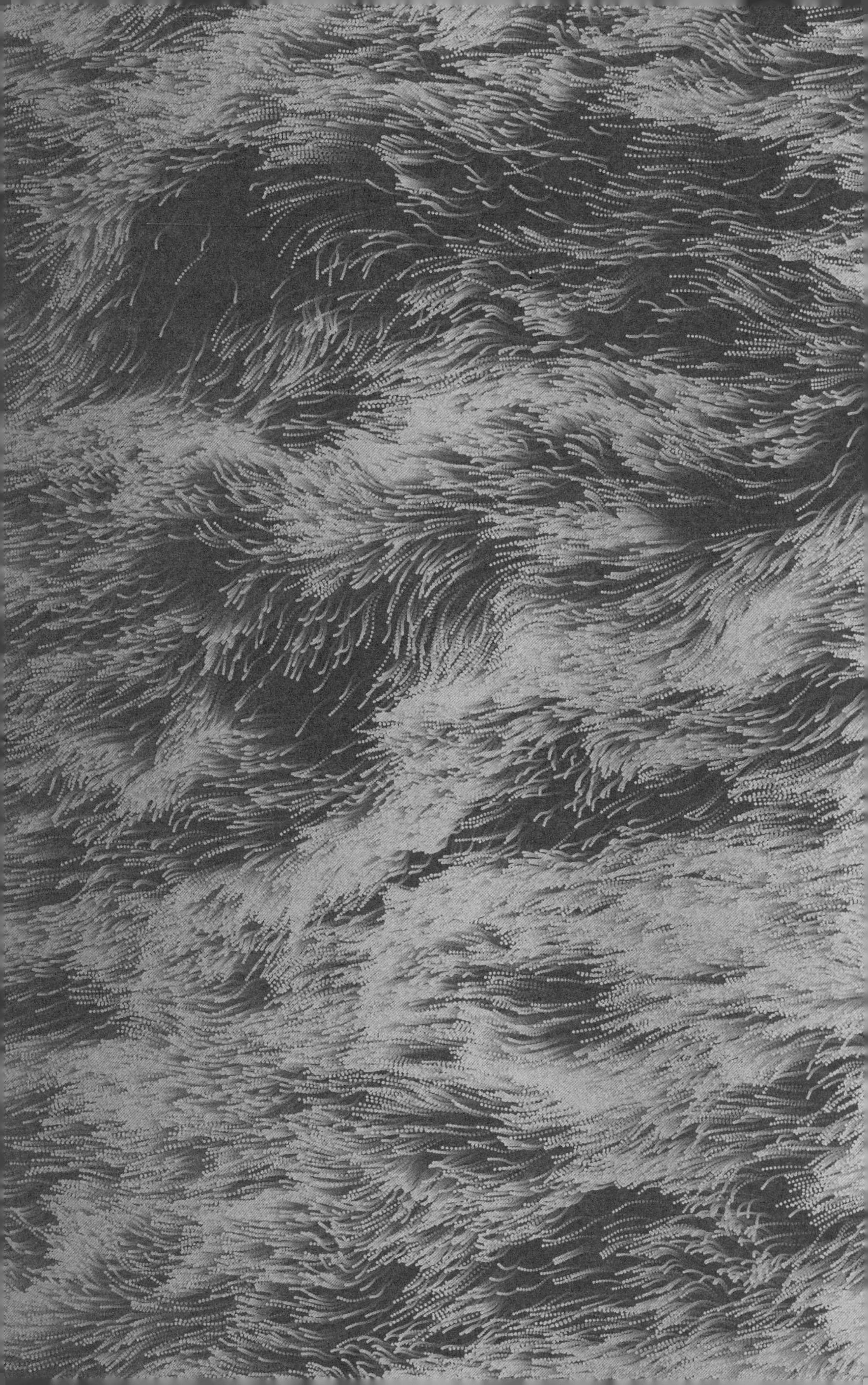